MECHANIZATION OF CULTIVATED CROPS

NIPA® GENX ELECTRONIC RESOURCES & SOLUTIONS P. LTD.
New Delhi-110 034

MECHANIZATION OF CULTIVATED CROPS

Surendra Singh
Ex. Professor of Agricultural Engineering
Department of Farm Machinery & Power Engineering
Punjab Agricultural University
Ludhiana - 141 004

Gursahib Singh Manes
Professor & Head
Department of Farm Machinery & Power Engineering
Punjab Agricultural University
Ludhiana - 141 004

and

Anoop Kumar Dixit
Associate Professor
Department of Farm Machinery & Power Engineering
Punjab Agricultural University
Ludhiana - 141 004

NIPA® GENX ELECTRONIC RESOURCES & SOLUTIONS P. LTD.
New Delhi-110 034

NIPA® GENX ELECTRONIC RESOURCES & SOLUTIONS P. LTD.

101,103, Vikas Surya Plaza, CU Block
L.S.C. Market, Pitam Pura, New Delhi-110 034
Ph : +91 11 27341616, 27341717, 27341718
E-mail: newindiapublishingagency@gmail.com
web: www.nipabooks.com

For customer assistance, please contact
Phone: + 91-11-27 34 17 17
Fax: + 91-11- 27 34 16 16
E-Mail: feedbacks@nipabooks.com

ISBN: 978-81-19215-93-5

Composed and Designed by NIPA®.

Preface

The first requirement of a farm machine is that it should be able to perform its intended function satisfactorily. The application of machines and better power sources to enhance agricultural production has been one of the most significant developments in world agriculture during the twentieth century. The results of this development can be seen in many aspects such as reduction in burden and drudgery of farm work and worker, increase in productivity and production per worker, precision in application of crop inputs, increase in cropping intensity due to timeliness of operations, increase in the quality of produce, reduction in grain losses and increase in farm employment. Mechanization is particularly advantageous when it can minimize a high peak labour demand that occurs over a relatively short period of time each year. Mechanization also encourages better management of farm inputs, improvement in working conditions and performance of jobs that would otherwise be difficult by hand. It also helps in reducing the cost of production.

This book **'Mechanization of Cultivated Crops'** covers the farm tools and equipment used in different operations such as land development, tillage, seeding/planting, interculture, fertilizer application, plant protection, harvesting and threshing and residue management. List of literature cited has been given at the end of the each chapter for the easy reference of the readers. This book has been designed as a textbook for the students of Farm Machinery at graduate and under Graduate levels. This book will also be helpful to the under graduate students of agriculture.

The authors wishes to express their appreciation to many individuals, organizations, agricultural machinery manufacturers and scientists working in the area of Farm Power & Machinery whose material has been used in the preparation of this textbook. Surely, there is a scope to improve the present compilation. The authors shall welcome the constructive criticism from the readers for future improvement.

Surendra Singh
Gursahib Singh Manes
Anoop Kumar Dixit

Contents

1 Introduction

The Indian agriculture is characterized by agro-ecological diversities in soil, rainfall, temperature and cropping system. Besides favorable solar energy, the country receives about three trillion cubic meter of water from rainfall. Of the total geographical area of 329 million hectares the net sown area has remained at about 140-142 million hectares during the last 40 years. The gross cropped area however has increased to about 200 million hectare in 2011-2012 (Anonymous, 2013). The total operational holding is estimated at 105 million with average size of the holdings of 1.6 ha. The Indian agriculture is not only characterized by small fragmented land, but also hill farming and shifting cultivation are also practiced. Out of an estimated 142 m ha net cultivated area around 40% is irrigated and rest is rain fed. About 65% of the population depend upon agriculture for their livelihood. The productivity of rain fed areas is very low as compared to the irrigated areas. The contribution of the rain fed areas to the overall production is about 44% of total farm output. One of the biggest challenges before the agricultural sector of India is to meet the growing demand of food grains to feed increasing population. This will require both higher energy inputs and better management of food production systems. Indian farmers have adopted improved seeds, fertilizer, plant protection chemicals, irrigation water and mechanization inputs which have helped them in increasing the food grain production.

Cereal crops or grains are mostly grasses cultivated for their edible grains or seeds. Cereal grains are grown in greater quantities and provide more energy worldwide than any other type of crop; they are therefore called staple crops. They are also a rich source of carbohydrate. In some developing nations, grain constitutes practically the entire diet of poor people. In developed nations, cereal consumption is more moderate but still substantial. Food grain production in India has increased from about 50.83 million tonnes in 1950-51 to over 257.44 million tonnes in 2011-12 (Table 1.1). Productivity of food grains increased from 522 kg/ha in 1950-51 to 2059 kg/ha in 2011-12. The net area under irrigation increased from about 18.1% in 1950-51 to over 49.3% of total cropped area now; gross irrigated area has also increased by over 300%. Groundwater irrigation has played the lead role in bringing more area under irrigation.

Rice and wheat are the major cereal crops of India. The well-known rice-wheat rotation covers about 37% of total area of India. This cropping system is extensively adopted in Northern India especially in Punjab, Haryana and western Uttar Pradesh. Most of the farmers prefer this crop rotation due to less farm activity required for growing the rice and wheat crops. Table 1.1 shows the production of food grain since 1950-51 onward. The production of rice was 20.58 million tonne in 1950-51 and increased to 87.1 million tonne in 2011-12. The increase in production of rice was mainly on account of kharif season output growth. The wheat production was also 93.90 million tonne in 2011-12. Similarly, total pulse production was 17.21 million tonne in 2011-12.

Table 1.1: Agricultural production – food grains.

(million tonnes)

Year	Cereals					Total Food grains
	Rice	Wheat	Cereals Coarse	Total Cereals	Pulses	
1950-51	20.58	6.46	15.38	42.42	8.41	50.83
1960-61	34.58	11.00	23.74	69.32	12.70	82.02
1970-71	42.22	23.83	30.55	96.60	11.82	108.42
1980-81	53.63	36.31	29.02	118.96	10.63	129.59
1990-91	74.29	55.14	32.70	162.13	14.26	176.39
2000-01	84.98	69.68	31.08	185.74	11.07	196.81
2005-06	91.79	69.35	34.06	195.20	13.39	208.59
2006-07	93.35	75.81	33.92	203.08	14.20	217.28
2007-08	96.69	78.57	40.76	216.02	14.76	230.78
2008-09	99.15	80.58	40.04	219.77	14.57	234.47
2009-10	89.09	80.80	33.55	203.44	14.66	218.11
2010-11	95.33	86.87	42.22	224.42	18.24	244.49
2011-12	87.10	93.90	30.42	211.42	17.21	257.44

Source: Anonymous (2013)

Oilseed crops have been the backbone of agricultural economy of India from time immemorial. Total area under these crops was about 16.64 million ha, with total production of 9.63 million tonnes in 1970-71 and increased to 27.22 million ha and 32.48 million tonne in 2010-11 respectively (Table 1.2). On the oilseed map of the world, India occupies a prominent position, both in regard to acreage and production. The important oilseed crops grown in this country in order of importance are soybean, groundnut, rapseed and mustard, sesame, linseed, safflower, castor, sunflower and niger. The oil content of the groundnut seed varies from 44 to 50%, depending on the varieties and agronomic conditions. Groundnut oil is edible oil. It finds extensive use as a cooking medium both as refined oil and Vanaspati ghee. It is also used in soap making, and manufacturing cosmetics and lubricants. Kernels are also eaten raw, roasted or sweetened. They are rich in protein and vitamins A, B and some members of B_2 group. Their calorific value is 349/100 g. The major groundnut-producing countries of the world are India, China, Nigeria, Senegal, Sudan, Burma and the USA. Out of the total area of 18.9 million ha and the total production of 17.8 million tonnes in the world, these countries account for 69% of the area and 70% of the production. In India about 70% of the area and 75% of the production are concentrated in the four states of Gujarat, Andhra Pradesh, Tamil Nadu and Karnataka.

Rapeseed and mustard yield the most important edible oil content of the seeds of different ranges from 30 to 48%. In the case of white mustard, the oil content ranges from 25 to 33%. The oil obtained is the main cooking medium in northern India and cannot be replaced by any other edible oil. The seed and oil are used as a condiment in the preparation of pickles and for flavouring curries and vegetables. The oilcake is mostly used as cattle feed. The leaves of young plants are used as a green vegetable. India occupies the first position, both with regard to acreage and production of rapeseed and mustard in the world. In India these crops occupy the second largest position after groundnut. The rape seeds and mustard producing states are Rajasthan, Uttar Pradesh, Punjab, Haryana, Assam, Bihar, Madhya Pradesh, West Bengal and Odisha.

Table 1.2: Area, production of oilseed crops in India.

Area-million ha; Production-million tonnes

Year	Oilseeds							
	Groundnut		Rapeseed & Mustard		Soybean		Total Oilseeds	
	Area	Production	Area	Production	Area	Production	Area	Production
1970-71	7.33	6.11	3.32	1.98	0.03	0.01	16.64	9.63
1980-81	6.80	5.01	4.11	2.30	0.61	0.44	17.60	9.37
1990-91	8.31	7.51	5.78	5.23	2.56	2.60	24.15	18.61
2000-01	6.56	6.41	4.48	4.19	6.42	5.28	22.77	18.44
2005-06	6.74	7.99	7.28	8.13	7.71	8.27	27.86	27.98
2006-07	5.61	4.86	6.79	7.44	8.32	8.85	26.51	24.29
2007-08	6.41	9.18	5.75	5.83	8.88	10.97	26.54	29.76
2008-09	6.16	7.17	6.30	7.20	9.51	9.91	27.56	27.72
2009-10	5.48	5.43	5.59	6.60	9.73	9.96	25.96	24.88
2010-11	5.86	8.26	6.51	8.17	9.55	12.73	27.22	32.48

Source: Anonymous (2013)

The sesamum seed is a rich source of edible oil. Its oil content generally varies from 46 to 52%. The grains may be eaten fried, mixed with sugar or as sweetmeats. Sesamum oil is used as a cooking-oil in southern India. It is also used for anointing the body, for manufacturing perfumed oils and for medicinal purposes. Sesamum-cake is a rich source of protein, carbohydrates and mineral nutrients, such as calcium and phosphorus. India, China, Sudan, Mexico, Turkey, Burma and Pakistan are the important sesamum producing countries. India ranks first, both in the area and production of sesamum in the world. Among the other states only Karnataka has a sizable area under sesamum. In the remaining states it is grown only on a small area and hence is a very minor crop there. Linseed occupies a greater importance among oilseeds; owing to its various uses and special qualities. In India, it is grown mainly for seeds, used for extracting oil. The oil content of the seed varies from 33 to 47%. Linseed oil is an excellent drying oil used in manufacturing paint and varnishes, oilcloth, waterproof fabrics and linoleum and as edible oil in some areas. Linseed-cake is a very good manure and animal feed. Linseed straw produces fibre of good quality. Linseed is used in making paper and plastics. The major linseed-growing countries are Argentina, Russia, India, U.S.A., Canada, Pakistan and Australia. India accounts for about 1.9 million ha, with a seed production of 0.5 million tonnes and occupies the third rank among the linseed-producing countries. In India, Madhya Pradesh leads in yield and acreage, followed by Uttar Pradesh, and Maharashtra. Madhya Pradesh and Uttar Pradesh together contribute to the national linseed production to the extent of about 70%.

Castor is generally grown for its, oil-yielding seeds. The oil content of the seeds varies from 35-58% in different varieties, the average being about 47%. Castor oil is being used widely for various purposes. It is used as a lubricant in high-speed engines and aero planes, in the manufacture of soaps, transparent paper, printing inks, varnishes, linoleum and plasticizers. It is also used for medicinal and lighting purposes. The cake is used as manure and plant stalks as fuel or as thatching material or for preparing paper-pulp. In the silk-producing areas, leaves are fed to the silkworms. The main castor-growing countries are Brazil, India, Russia and Argentina. India ranks first in area (4,48,000 ha) but second in production (1,40,000 tonnes). Andhra Pradesh (67.2%), Gujarat (12.7%), Karnataka (7.1%) and Odisha (5.8%) account for over 90% of the area and also production.

Safflower crop is now cultivated, primarily for its seeds which yield oil, though at one time it used to be grown for the extraction of a dye also. The seeds are edible and are eaten after roasting. Their oil content varies from 24 to 36%, depending on the variety, soil, climate and other conditions. The important safflower growing countries, besides India, are the USA, Mexico, Ethopia, Spain, Australia and Russia. In India, it occupies 590,000 ha with a production of nearly 1,30,000 tonnes. Over 70% of the area is concentrated in the states of Maharashtra (40.4%), Karnataka (26.0%) and Andhra Pradesh (8.0%). Sunflower is mainly grown for its oil. The oil is used for culinary purposes, in preparation of vanaspati and in the manufacture of soaps and cosmetics. It is especially recommended for heart patients. Its cake is rich in protein and is used as a cattle and poultry feed. The major sunflower-producing countries of the world are the Russia, Argentina, Bulgaria, Pumania, Turkey and South America. Sunflower, as an oilseed crop, was introduced into India in 1969. Niger crop is grown for its seed, used for extracting oil, which is about 37 to 43% of the seed weight. The oil is used for culinary purposes, for anointing the body, for manufacturing paints and soft soaps, for lighting and lubrication and for manufacturing cosmetics. The oilcake is a well-known cattle feed. India is considered to be the chief Niger-producing country in the world, with an area of 0.48 million ha, and with an annual estimated production of 0.12 million tonnes of seed. It is mainly grown in the states of Madhya Pradesh, Bihar, Maharashtra, Odisha and Tamil Nadu, of which the largest area of 0.238 million ha is in Madhya Pradesh.

The total sugarcane area during 1984-85 was 2.95 million ha, which rose to 4.94 m ha by 2010-11 (Table 1.3). Sugarcane production also had similar increase and during the year 2010-11, it was an all time high of 342.38 million tonnes. Sugarcane is cultivated in the most of the Indian states at present. Uttar Pradesh has the largest area almost 50% of the cane area in the country, followed by Maharashtra, Karnataka, Tamil Nadu, Andhra Pradesh, Gujarat, Bihar, Haryana and Punjab. These nine states are the most important sugarcane producing states. Sugarcane production is highest in Uttar Pradesh followed by Maharashtra. Productivity-wise, Tamil Nadu stands first with over 100 tonnes/ha followed by Karnataka and Maharashtra.

Table 1.3: All India area, production and productivity of sugarcane.

Year	Area(million ha)	Production(million tonnes)	Productivity(kg/ha)
1950-51	1.71	57.05	33362.57
1960-61	2.42	110.00	45454.55
1970-71	2.62	126.37	48232.82
1980-81	2.67	154.25	57771.54
1984-85	2.95	170.32	57735.59
1990-91	3.69	241.05	65325.20
1994-95	3.87	275.54	71198.97
2000-01	4.32	295.96	68509.26
2005-06	4.20	281.17	66945.24
2006-07	5.15	355.52	69033.01
2007-08	5.06	348.19	68812.25
2008-09	4.42	285.06	64493.21
2009-10	4.17	292.30	70095.92
2010-11	4.94	342.38	69307.69

Source: Anonymous (2013)

There are approximately 400 vegetable crops (including root and tuber crops) that are commercially cultivated worldwide (Kays and Silva Dias, 1996). Root crops consist of beets and carrots, and tuber crops potatoes and sweet potatoes and the leaves of root crops, such as beet tops. Uttar Pradesh is the first leading producer with 60% for fresh market and 40% for processing. Potato is most widely grown vegetable crop in the country with a share of 25.7%. The area under potato cultivation is 1.86 million ha with total production of 42.34 million tonnes (Table 1.4). The main varieties of potato grown in the country are Kufri, Chandramukhi, Kufri Jyoti, Kufri Badshah, Kufri Himalani, Kufri Sindhuri, Kufri Lalima etc. Uttar Pradesh is the leading potato growing state in the country followed by West Bengal and Bihar.

Table 1.4: All India area, production and yield of potato and sweet potato.

Year	Area(million ha)	Production(million tonnes)	Productivity(kg/ha)
1990-91	0.94	15.21	16254.00
2000-01	1.22	22.49	18404.00
2001-02	1.21	23.92	19806.00
2002-03	1.35	23.27	17300.00
2003-04	1.29	23.06	17887.00
2004-05	1.32	23.63	17923.00
2005-06	1.40	23.91	17058.00
2006-07	1.48	22.18	14943.00
2007-08	1.55	28.47	18331.00
2008-09	1.83	34.39	18810.00
2009-10	1.84	36.58	19951.00
2010-11	1.86	42.34	22763.44

Source: Anonymous (2013)

Engineering inputs are vital for modernization of agriculture, agro-processing, and rural living. Mechanization of agriculture is needed for development and optimal utilization of natural resources leading to higher productivity and reduced cost of production for greater profitability, economic competitiveness and sustainability. Mechanization also imparts capacity to the farmers to carry out farm operations with dignity, comfort and freedom from drudgery, making the farming agreeable vocation for educated youth as well. It helps the farmers to achieve timeliness in farm operations and apply costly input with reduced quantity for better efficacy and efficiency. Small and marginal farmers can now make use of high capacity agricultural machines on custom hire basis. Modernization of production and post-production agriculture utilizing a proper blend of conventional and renewable energy sources facilitate the application of cutting edge agricultural engineering technologies for food, nutrition and environmental security to the ever increasing Indian population and also to that of the world.

Intensive cultivation as a result of introduction of high yielding varieties in mid 1960's required higher energy inputs and better management practices. Land preparation, harvesting, threshing and irrigations are the operations, which utilize most of the energy used in agriculture. For desired cropping intensity with timeliness in field operations animate energy sources are no longer adequate. Farmers are encouraged to adopt mechanical power sources to supplement animate power; as a result, the share of animate power in agriculture has decreased from 92% in 1950-51 to 8% in 2011-12. Irrigation, tillage and threshing operations, which not only require more energy but also are arduous to perform, have been mechanized. Subsequent chapters deals with various tools and equipment used presently in cultivation of different crops grown in the country

References

Anonymous. 2013. www.indiastat.com

Kays, SJ., Silva Dias, J.C. 1996. Cultivated vegetables of the world. Exon Press. Athens. GA.

2 Land Leveling Equipment

Introduction

The work on field drains and farm ditches are essential for providing proper irrigation to the crop. Leveling of field is equally important for smooth operation of various field activities including irrigation. Land grading in irrigated agriculture helps in uniform application of water, better water regulation and saving in irrigation time. Land grading helps in conservation of soil and moisture under rainfed conditions. It provides desired surface drainage under irrigated and rainfed conditions. Land leveling helps in the smooth operation of sowing/ planting and harvesting equipment.

Land leveling operations required for an area depends upon the topography, soil type, soil depth, prevailing land slope, rainfall characteristics, crops to be grown, source of water supply, method of irrigation and other special features of the site including the preference of farmer. Based on the initial topography of land the layout of individual fields, irrigation and drainage systems are planned. Usually the field is not graded to a truly level surface, but a gentle uniform slope is maintained to meet the requirements of irrigation and drainage.

Land leveling operations may be grouped into three parts viz. rough grading, land leveling and land smoothing. Rough grading is the removal of mounds, dunes and other irregularities on the land surface. It also includes filling of pits, depressions and gullies. Land leveling reshapes the land surface to a planned grade. Land leveling requires movement of large quantities of soil over a considerable distance. Land soothing is done prior to seeding as a regular land preparation practice. Planning and survey is done prior to land-leveling operation. The entire area should be taken into consideration for the purpose of planning and land leveling operation.

The basic methods for land leveling are plane method, profile method, plan inspection method and contour adjustment method. Each method has advantages and disadvantages and is best adapted to specific site conditions. In the plane method, the centroid of area is found and plane is passed through this point at an elevation equal to average elevation of field. Under this condition, irrespective of slope of plane, the volume of excavation equals the volume of fill. In profile method, ground profiles are plotted along the grid lines using grid point elevation. Final land profiles are chosen by trial and error method such that cuts and fills balance. This method is suitable for leveling flat lands or lands with undulating topography. In this method, profiles are commonly plotted in one direction and datum lines are kept at proposed profiles. Two-way profiles are used when slopes are given in both directions. In plan inspection method, a suitable down slope and cross slope is assumed and cuts and fills are calculated. Assumed slopes are altered until a balance of cuts and fills are obtained. In contour adjustment method, a contour map of area is drawn using the levels. The ground surface expected after grading is shown on the same map with new contour lines such that a uniform desired slope is obtained. The cuts and fills are estimated at the grid points by interpolating between contour lines and by taking the difference in elevations between original and new surface.

Uneven soil surface has a major impact on the germination, stand, and yield of crops due to inhomogeneous water distribution and soil moisture. Therefore, land leveling is a precursor to good agronomic, soil, and crop management practices. In irrigated areas or even otherwise land levelling is an important farm operation. It is used to level the field to ease not only field operation by machines but receives uniform level of irrigation water. The farmers commonly use wooden logs or planks as levellers. Traditionally farmers level their fields using animal drawn (Fig. 2.1) or tractor-drawn levelers (Fig. 2.2 and Fig. 2.3), Singh (2007). It is seen that even the best leveled fields using traditional land leveling practices are not precisely leveled and this leads to uneven distribution of irrigation water. The common practices of irrigation in intensively cultivated irrigated areas are flood basin and check basin irrigation systems. These practices on traditionally leveled or unleveled lands lead to water logging conditions in low-lying areas and soil water deficit at higher spots.

Fig. 2.1: Land leveling by bullock-drawn planker (leveler).

Fig. 2.2: Land leveling by tractor operated planker (leveler).

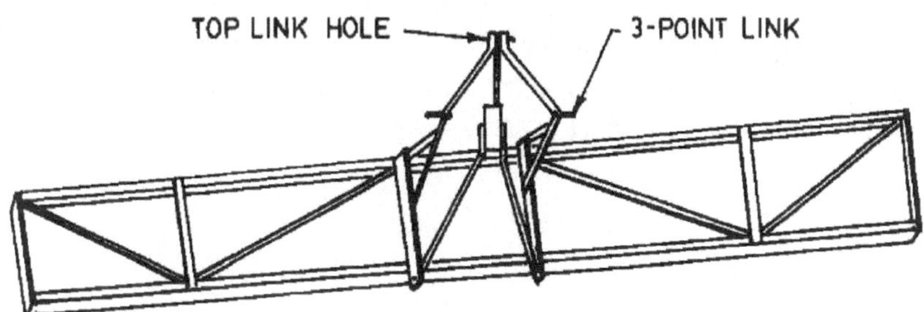

Fig. 2.3: Tractor operated land leveler also called float or clod crusher or planker

Soil Scoop: It is used in excavating ditches, cleaning drains and moving soil over short distances. It is available both animal-drawn and tractor operated (Singh, 2007). It consists of blade, soil trough, hitching-loop and a handle (Fig. 2.4). The angle of cutting blade varies from 12-15 degrees. The blade is bolted to the soil trough made of mild steel sheet.

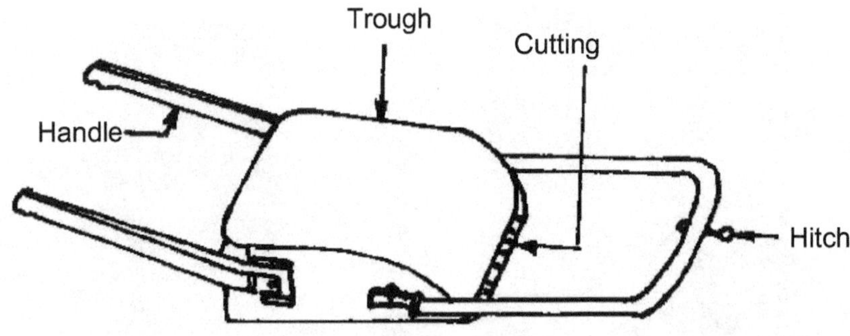

Fig. 2.4: Soil scoop.

Buck Scraper: The other improved type of land leveller generally used on large farms is called levelling 'Karaha' or scraper (Fig. 2.5), Singh (2007). It is suitable for land preparation operations such as scraping, grading, leveling and back filling. It is also used for irrigation, terrace work and general cleaning of field. It is available in animal-drawn and tractor-drawn. It consists of hitch system, replaceable cutting blade with sharp edge, and a curved plate with side wings, which form a bucket. The blade is made from medium carbon steel or low alloy steel, hardened and tempered to suitable hardness. During operation, the blade digs into the soil and extra soil is collected in the bucket, which is released in the depressions of the field. The angle and pitch of leveler is adjustable. The leveler can also angled left or right, or reversed for back filling. The amount of work done depends on the various factors such as hardness of soil, transportation distance and volume of soil cut each time. It works well if haulage distance does not exceed 50 m. For long distances tractor operated pull type scraper can be used (Fig. 2.6).

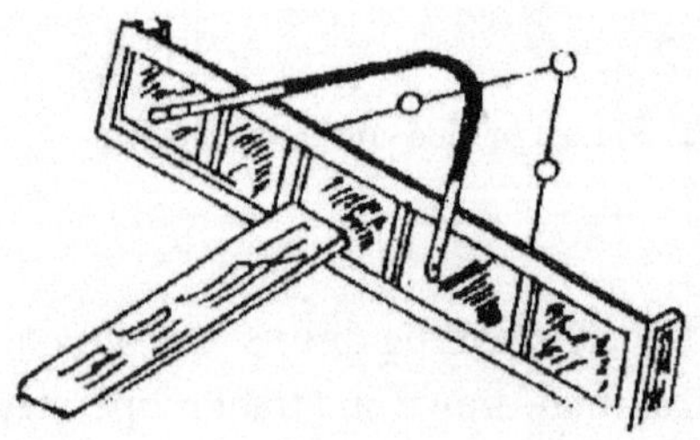

Fig. 2.5: Animal-operated buck scraper.

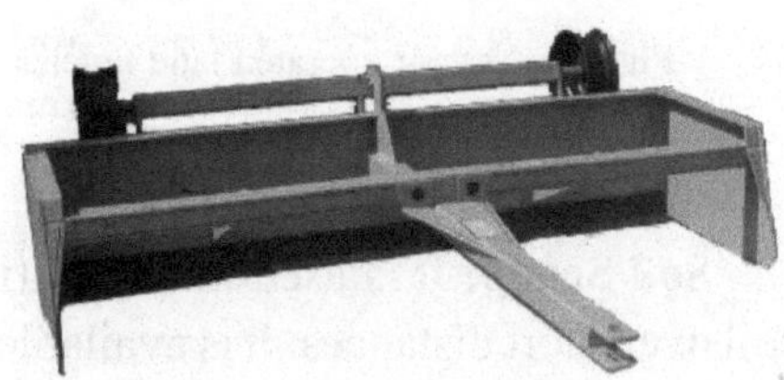

Fig. 2.6: Tractor-operated mounted/pull-type scrapers

Tractor operated terracer blade: The terracer blade is used for grading, levelling of fields, filling of depressions and smoothening of field for irrigation. It is attached to the tractor with the 3-point linkage system and is hydraulically controlled (Singh, 2007). It consists of replaceable blade attached to curved steel body, side wings and indexing arrangement for tilting and angling of the blade (Fig. 2.7). The cutting blade is made of medium carbon or alloy steel and the cutting edge is hardened and sharpened. The blade with body is also called mould board. The mould board can be angled left or right by lifting the spring loaded latch pin and by turning the mould board. To tilt the blade for ditching or terracing the blade is tilted to the desired angle by moving the index pin. The depth of cutting is controlled by hydraulic system of the tractor. The blade can be pitched forward and back or tilted at 15^0 to 30^0 left or right. It can be reversed for back filling. They can be operated by 35 – 45 hp tractor. Working capacity of the blade is 0.3 to 0.4 ha/h.

Fig. 2.7: Different views of tractor operated terrace blade.

Hydraulic Scraper: It is used for collecting the soil from one place and unloading at the other. It is used for rough leveling, cutting of high spots and filling of depressions. It is operated by 50 hp tractor and above (Singh, 2007). The hydraulic scraper is towed behind the tractor (Fig. 2.8a). Self-propelled scrapers are also available (Fig. 2.8b). The scraper consists of cutting blade, hydraulic system, hitch point, hitch bar, apron, bowl, wheels, apron cylinder, side frame, bucket cylinder, spring, and side arm. The scraper working is controlled by the hydraulic arrangement. The blade is made of alloy steel and has self-sharpening tungsten carbide cutting edge. For operation, the scraper is attached to the tractor, hydraulic system connected and apron is raised. With the forward movement of the tractor, the blade penetrates into the soil and bucket bowl gets filled. The apron is closed after the bucket is filled and the scraper is moved to the point of unloading. For unloading, the bucket is tilted hydraulically. The working capacity of scraper is 2.5 to 3.5 m³/h. The scrapers are cheaper for moving large amount of earth. But it cannot be used for long distances. It takes wide turning circle and cannot be operated on steep slopes. It can climb only on gentle slopes. It cannot cut vertical or near vertical land or rock. Cutting blade width is 1500 – 2100 mm.

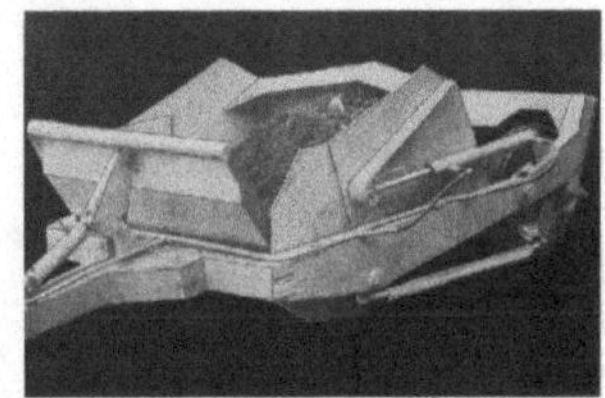

a) Tractor operated hydraulic scraper

b) Self-powered scraper

Fig. 2.8: A view of hydraulic scraper.

Laser Land Leveler: Laser leveling of agricultural land is a recent resource-conservation technology. It has the potential to change the way food is produced by enhancing resource-use efficiency of critical inputs without any disturbing and harmful effects on the productive resilience of the ecosystem. In spite of several direct and indirect benefits derived from laser land leveling technology, it is yet to become a popular farming practice in the developing and the underdeveloped countries. For accelerating its popularization and large-scale adoption, it requires a number of well-considered and synchronized research, extension, participatory, economic and policy initiatives keeping in view the long-term sustainability of our production systems. Popularization of this technology among farmers in a participatory mode on a comprehensive scale, therefore, needs appropriately focused attention on priority basis along with requisite support from researchers and planners. The change in our vision of future agriculture in relation to food and nutritional security, environmental safety and globalization of markets demands improving resource-use efficiency considerably to reach the desired growth levels in food production and agricultural productivity. Laser leveling is evidently one of the ways by which we can address these issues to a great extent.

Declining water table and degrading soil health are the major concerns for the present agriculture. The enhancement of water use efficiency and farm productivity at field level is one of the best options to redress the problem of water scarcity. Laser land levelling is one such technology which helps in using water efficiently, reduced irrigation time and enhanced productivity not only of water but also of other farm inputs and also reduces nutrient losses through uneven irrigation in undulating fields. Laser land leveling is a laser guided precision leveling technique used for achieving very fine leveling with desired grade on the agricultural field.

Laser land leveller is trailed type equipment used for achieving precise levelling with desired grade (Sidhu *et al.*, 2007; Anonymous, 2008 and 2012). This two meter wide automatic levelling operation can be successfully carried with 45 hp or above tractor. Components of laser land leveler include laser emitter, laser beam receiver, control box, hydraulic valve and laser eye (Fig. 2.9). Laser levelling uses a laser transmitter unit that constantly emits 360° rotating beam parallel to the required field plane. This is received by a laser receiver (receiving unit) fitted on a mast on the scraper unit. The signal received is converted into cut and fill level adjustment and the corresponding changes in scraper level are carried out automatically by a two way hydraulic control valve. Laser leveling maintains the grade by automatically performing the cutting and filling operation. Both level grade and slope grade (one way or two ways) can be achieved with the help of this precision equipment. The field is cultivated and planked before using the laser land leveler. A grid survey is performed using grade rod to identify highs and lows in the field and mean grade is found. A grid spacing of 10 m x 10 m is maintained for accurate land survey; however this spacing can be varied depending upon the size of the field. For practical purposes and with experience, grid surveys can be done by pacing off the distances (rather than measuring). A map is then drawn to indicate which areas are high (and require soil to be cut) and the lows which require soil to be added. It can cover about 0.4 ha/h; however capacity depends on type of soil, topography of land and cut & fill.

The laser emitter unit sends continuous self leveled laser beam signal with 360° laser reference up to a command radius of 300-400 m (depending upon its range) for auto-guidance of the receiving unit. The laser emitter is mounted on a tripod stand placed just outside the field to be laser leveled and high enough to have unobstructed laser beam travel. Different working components & controls on the laser emitter unit includes laser emission indicator, low battery indicator, off/on power button, manual grade buttons, charge jack, battery assembly and manual mode indicator for setting of desired grade. The laser receiver mounted on the scraper is a unidirectional (360°) receiver that detects the position of the laser reference plane and transmits it to the control box mounted on the tractor. Further this control box directs the double actuating hydraulic valve for desired upward and downward movement of scraper blade to obtain the leveled field. The grade position LED's indicate the position of the machine's blade relative to the plane of the laser light from the laser emitter. These lamps function in the same way as the grade position lamps on the control box mounted on tractor except they flash rapidly instead of lighting solidly.

The control box is mounted on the tractor so that the operator can easily access the switches and view the indicator lamps. The control box has the main control unit for actuating the double acting hydraulic valves. The control box receives and processes signals from the bucket. It displays these signals to indicate the drag bucket's position relative to the finished grade. The control box is set to manual for initial adjustments of scraper blade before starting operation. When the control box is set to automatic position, it provides electrical output for driving the hydraulic valve to operate scraper automatically. The three control box switches are On/Off, Auto/Manual, and Manual Raise/Lower (which allows the operator to manually raising or lowering the drag bucket). The valve assembly regulates the flow of tractor hydraulic oil to the hydraulic cylinder to raise and lower the scraper blade. The oil supplied by the tractor's hydraulic pump is normally delivered at 2000-3000 psi pressure. As the hydraulic pump is a positive displacement pump and always pumping more oil than required, a pressure relief valve has also been provided in the system to return the excess oil to the tractor reservoir. The solenoid control valve controls the flow of oil to the hydraulic ram which raises and lowers the bucket. The desired rate at which the bucket could be raised and lowered is dependent on the operating speed.

The faster the ground speed, the faster the bucket will need to be actuated. The rate at which the bucket will raise and lower is dependent on the amount of oil supplied to the delivery line. Laser eye is mounted on the grade survey rod for obtaining the level of the field. It contains a laser receiving panel and when the laser emitted by the laser emitter panel falls in the center of this eye a continuous beep indicates the level of that specific point w.r.t. the laser emitter. The grade of that point is then read from grade rod.

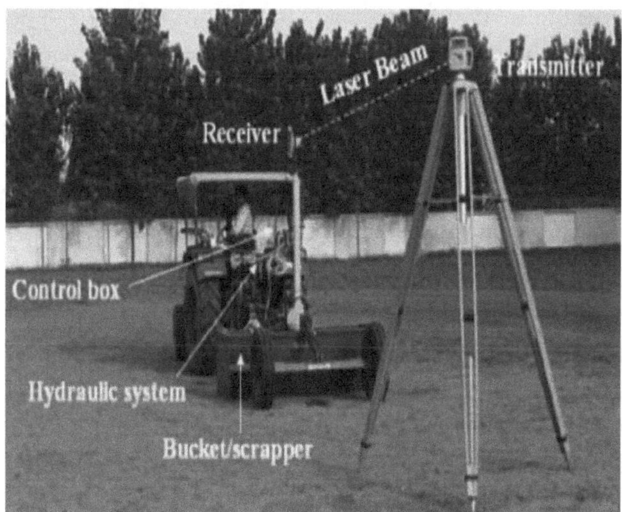

Fig. 2.9: Showing different components of Laser Levelling System

The system includes a laser-transmitting unit that emits an infrared beam of light that can travel up to 700–1000 m in a perfectly straight line. The second part of the laser system is a receiver that senses the infrared beam of light and converts it to an electrical signal. The electrical signal is directed by a control box to activate an electric hydraulic valve. Several times a second, this hydraulic valve raises and lowers the blade of a grader to keep it following the infrared beam. Laser leveling of a field is accomplished with a dual slope laser that automatically controls the blade of the land leveler to precisely grade the surface to eliminate all undulations tending to hold water. Laser transmitters create a reference plane over the work area by rotating the laser beam 360 degrees. The receiving system detects the beam and automatically guides the machine to maintain proper grade.

The fields can be leveled with this laser levelers to perfectly flat/ horizontal fields (level with zero grade) or to any single or dual slope (or grade) depending upon soil & water and agronomic practices. If the field is to be prepared as perfectly flat field i.e. no slope, the set up of the laser emitter is simple and does not need any alignment. Set the laser emitter for zero level grades (as per instructions in operator's manual) which will provide a level plane of laser beam in all directions. If the field is to be graded for single or dual slope, laser emitter requires its axes to be aligned accordingly (as per instruction in the operator's manual). The laser emitter will then provide a plane of laser beam at desired slope(s). Major soil movement should be done with traditional equipment or specific machine depending upon quantum of soil movement prior to usage of laser land leveler. Fix the laser emitter and laser eye on tripod and graded rod respectively. Adjust/align the emitter for level grading or sloped grading. Establish the level of the field using grade rod at different locations in field. While taking the level on grade rod the laser eye and the laser emitter should be in line and continuous beep should sound from laser eye after adjusting it up and down. Record the field levels at corresponding points selected in the field at every 10-15m depending upon the size of field. More the points selected for survey more will be the precision. Mark the points/locations where levels have been recorded with pegs. Calculate the average field level obtained after the field survey. Locate the point similar or nearest to the average level obtained.

The steps involved in accurate usage of laser land leveler in field are (i) calibration of equipment, (ii) setting-up of equipment for field operation (horizontal level or graded field i.e. one way/both ways), and level survey of field. It is essential to check the laser for calibration (as per guidelines provided in the operator's manual). While calibrating, the equipment should be placed on a flat/level surface of tripod approximately 30 m from the wall. Once the desired calibration has been set, press the power button at once to store the information and turn the power off. The equipment is set to the new calibration when power is re-applied. The equipment should be calibrated only when required. The following guidelines are followed for setting up the laser emitter, laser receiver, laser eye and control box on a tractor:

a) The laser emitter and laser eye are fully charged or equipped with replicable batteries with sufficient battery back-up before taken to field.

b) Check the electrical connections of control box.

c) Choose a location in the field for the laser emitter (to be fixed on tripod) where obstruction, such as trees and buildings, passages etc does not block the plane of laser light. The laser receiver on the scraper should be able to sense the plane of laser light all times.

d) As far as possible, set up the laser emitter and receiver at a height above the tractor's canopy or roll over protection attachment to avoid any blocking the plane of laser beam as the machine moves around the field.

e) Fix the laser eye on the graded rod for the field level survey.

After locating the average level of the field required for flat level or sloped grade, following steps should be followed for benching the equipment:

a) Set the scraper blade and laser beam receiver at the location where average/nearest level exists (Fig. 2.10).

b) Set the control knob/switch on control box mounted on tractor to manual. Then set the scraper blade just above the selected location using raise and lower switch/knob on the control box (Fig. 2.10).

c) After setting the scraper blade, adjust the laser beam receiver mounted on scraper mast at such as point where green light blinks on the control box indicating that laser beam emitter and laser receiver are in line. Set the control knob/switch from "Manual to Auto" and start operating the tractor & leveler as per survey map/sheet. The operator must take minimum time and soil to pick, carry and place the soil following the survey map/sheet.

2.10: A view of laser land leveler working in the field

The benefits of laser land levelling are:

(i) Enhances water application efficiency,

(ii) Saving in irrigation water up to 20-30%,

(iii)Better crop stand due to even application of fertilizers and other inputs resulting improvement in crop yield by 5 to 10%,

(iv) Improves weed control efficiency,

(v) Less area under bunds/channels increases up to 8 to 10% area under the crop,

(vi) Reduces labor requirement for irrigation,

(vii) Reduction in time and water for irrigation,

(viii) Uniform distribution of water,

(ix) Less water consumption in land preparation,

(x) Precise level and smoother soil surface,

(xi) Uniform moisture environment for crops,

(xii) Good germination and growth of crop,

(xiii) Uniformity in crop maturity,

(xiv) Reduced seed rate, fertilizers, chemicals and fuel requirements, and

(xv) Increase in field areas by about 3-5%.

References

Anonymous. 2008. Success Stories. AICRP on Farm Implements and Machinery, CIAE Bhopal. Extension Bulletin No.: CIAE/FIM/2008/80.

Anonymous. 2012. Directory of Successful Farm Machinery in SAARC Countries. SAARC Agriculture Centre. BARC Complex, Farmgate, Dhaka – 1215 Bangladesh.

Sidhu H S., Mahal J S., Dhaliwal I S., Bector Vishal, Singh Manjit, Sharda Ajay and Singh Thakur. 2007. Laser Land Leveling: A Boon for Sustaining Punjab Agriculture. Department of Farm Power and Machinery, Punjab Agricultural University, Ludhiana.

Singh Surendra. 2007. Farm Machinery – Principles and Applications. Directorate of Information & Publication of Agriculture, Indian Council of Agricultural Research, Krishi Anusandhan Bhawan-I, Pusa Campus, New Delhi.

3 Tillage Equipment

Tillage operations are the basic operation, which need to be performed at proper moisture and temperature condition (climatic factor) to attain high degree of germination and better plant growth. Tillage may be defined as mechanical manipulation of soil for the purpose of germination and plant growth but it also does the followings:

- develops a desirable soil structure for a seedbed i.e. provide adequate air supply and exchange within the soil and minimize resistance to seed penetration,
- prepares granular structure for rapid infiltration and retention of rainfall,
- controls weeds or removes unwanted plants,
- manages plant residue i. e. mixing of trash for better tilth and decomposition and reduction of soil erosion due to trash in upper layer,
- minimizes soil erosion by following contour tillage listing,
- incorporates fertilizer, pesticides, soil amendments etc,
- accomplishes segregation i.e. removal of rocks, foreign matters, roots etc,
- prepares land for sowing, planting and irrigation; and
- smooth or otherwise prepares the soil surface for harvesting operations.

The importance of optimizing tillage operation and improving tillage tool design can be gauged from the fact that more than 140 bm^3 (200 b ton) of soil need to be turned or stirred if ploughed once in India. A primary tillage operation is the major soil evolving operation designed to reduce soil strength, cover plant material and rearrange the aggregates. Secondary tillage operation creates refined soil condition. Plows and disk type implements cut straws and partially invert the soil.

Puddling of soil generally refers to breaking down soil aggregates at near saturation into ultimate soil particles and is one of the common operations in low land rice fields. It is normally done after initial ploughing and allowing about 50 to 100 mm of standing water in the field. However, in low land condition the farmers often flood the field prior to ploughing and puddling to weaken the mechanical strength of the soil. Retention of standing water on the rice field helps weed control and oxidation-reduction. Such conditions help achieving nutrient balance, and a soft soil suitable for transplanting rice seedlings. Puddling helps retain standing water in the rice field by producing fine soil particles making thin layer that reduce soil porosity, thus reducing percolation losses of water and nutrients. Puddling is also beneficial because it controls weeds, levels the soil surface and provides a homogenized puddled tilth.

Puddling is done with an animal, power tiller or tractor operated implement (puddler) such as ploughs, comb harrow, patela puddler, ladder puddler and rotary puddler. The degree of puddling is however, dependent on the type of puddler and on intensity of puddling. Rotary puddlers generally are better than ploughs because their rotary motion continuously changes the direction of the shear stress and therefore matches the weakest fracture plane within a clod. Further the rotary puddlers tend to compact the sub soil, chop and press down organic matter and require relatively low draft as compared to ploughs.

The early man learned by experience to till the soil for growing crops. Later on, the manpower was replaced by animal power for tilling. In about last one hundred years, tractors fueled by hydrocarbons have replaced draft animals as a source of farm power in developed countries and are fast replacing them in developing countries. Tillage is one of the major crop production operations and is an important contributor to the total cost of the crop production. It is a time and energy consuming operation. With the increasing emphasis on timeliness of planting operations for higher yield and energy conservation, the necessity and extent of tillage operation has come under wide investigation.

Tillage Equipment

Animal drawn: Tillage has been traditionally done by bullock-drawn soil stirring ploughs (wooden ploughs and steel ploughs (Fig. 3.1). It is most commonly used implement for land preparation. It is also called 'Desi' plough. It consists of body, shoe, share, beam and handle. The body is the main part of plough to which shoe, beam and handle are generally attached. The share is working part of plough and is attached to shoe with which it penetrates into the soil. The shoe also supports and stabilizes the plough at a required depth. The beam is generally a long wooden piece that connects main body of plough to the yoke. A wooden piece called handle is attached vertically to the body to control the plough. It cuts the soil to a depth of 5-10 cm but does not invert the soil. The field capacity of this plough is 0.2-0.3 ha per day and its draft requirement is about 50-70 kg. To plough one hectare of land a man has to walk about 68 km.

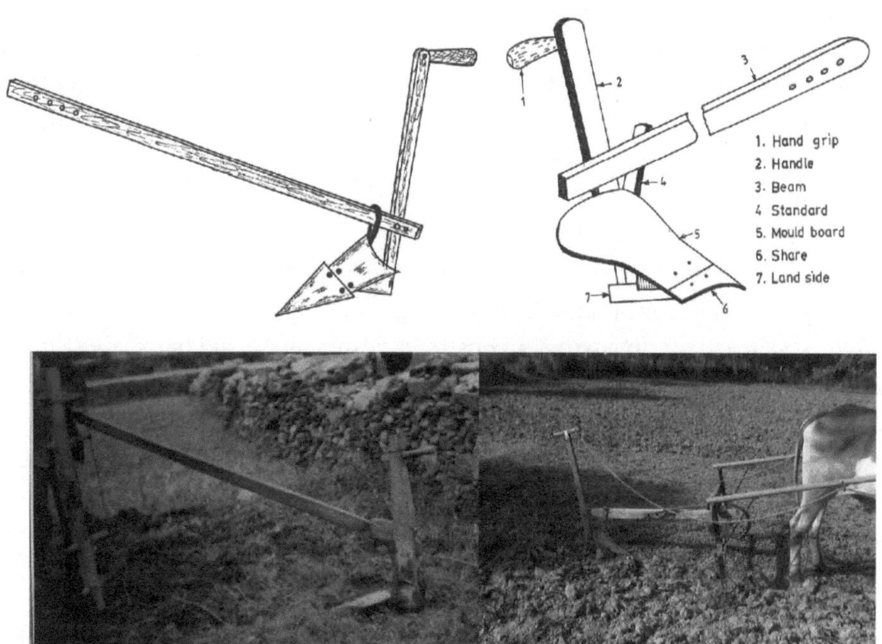

Fig. 3.1: Bullock-drawn soil stirring ploughs (wooden ploughs and steel ploughs).

Bullock-drawn mould board plough is very useful for medium and heavy soils for first ploughing for those fields having stubble and invested with weeds. The plough is available in different sizes, ranging from 15 to 20 cm in width and can easily be operated with a pair of bullocks. There are two types of animal-drawn ploughs; viz. one or two way ploughs and left hand or right hand ploughs (Singh, 2007). Most of the walking type mould board ploughs are one way ploughs, that is they are designed to throw the furrow slice to only one side in the direction of motion. Two ways ploughs or Turn Wrest ploughs are suitable for terraced land of hilly tracts and have the advantage that they do not upset the slope of the land nor leave dead or back furrows in the middle of narrow fields. This is because the two bottoms are used alternatively and the furrow slices are thrown on the same side. Some of the two-way ploughs have a single bottom, but the provision is made to change the direction of throw of furrow slice at the end of the plot where the bullocks turn. Most of the mould board ploughs are right hand ploughs, throwing furrow slice to right. Left-hand type ploughs are rare in India because bullocks are trained to take turn on their left while ploughing. Mould boards are generally made of high carbon steel.

The main parts of mould board plough are share, mould board, landside, frog, beam and handle (Fig. 3.1). The share is the part of the plough that penetrates into soil, cuts the furrow slice and passes it onto the mould board. It takes the greatest amount of wear. It is fastened to the frog with the help of countersunk bolt to keep the top surface smooth. There are five main parts of the share that play important role in stabilizing the plough bottom on the ground and in cutting the furrow slice. *Share point* enters first in the soil and supports the plough bottom. *Cutting edge* also called throat of the share cuts the furrow slice from main soil body. *Wing of the share* supports the plough bottom. *Gunnel* of the share supports the plough bottom against the furrow wall. *Cleavage* forms joint between mould board and share on the frog. There are different types of share viz. Slip share, slip nose share, shin share and bar share (Singh, 2007). The share of plough is commonly made of high carbon steel or chilled cast iron. The steel mainly contains 0.7-0.8% carbon and 0.5-0.8% manganese besides other minor elements. There are self-sharpening shares with a hard alloy surface along the share blade that always remain sharp despite wear. The blade sharpens itself in all soils except sand and stone. These shares do not require periodic forging and have much longer life as compared to conventional shares. When share point is blunt by 3-4 mm, the plough draft increases by about 25 per cent depending upon type and moisture content of the soil. The working side of share should be sharpened at an angle of 25-40 degrees.

Slip share is a common type of share used by the manufacturers. It is one piece share with curved cutting edge and entire share has to be replaced if it is worn out. In a slip nose, share point of share is provided with a small detachable piece and can be replaced as and when required. In a shin share, shin is an additional part. In a bar share, point of share is provided with an adjustable bar. Full-cut shares are used for high speed and slat bottom where soil is having heavy root crop. Narrow-cut shares are used for soils where roots are not the problem, penetrate better and pull lighter than full-cut shares. Heavy-duty deep suck shares are used in abrasive, rocky and hard soil. Hard surfaced shares are recommended for extremely abrasive soils where regular shares wear out quickly. Chilled cast shares are used in light, abrasive, sandy and gravely soils where good scouring is not a problem.

Landside is the part of the plough that slides along the face of furrow wall. It helps to resist the side pressure exerted by furrow slice on the mould board. It helps in stabilizing the plough while it is in operation. Frog is the part of plough bottom to which the share, mould board and landslide are attached rigidly. It is an irregular piece of metal and made of cast iron and to which the different parts of the plough are attached. Plough bottom is of different types depending upon the soil type, condition and moisture content. The main functions of a plough bottom are i) to cut the furrow slice loose, ii) to loosen or pulverize the soil, iii) to turn the furrow slice over to the desired angle (50°), and iv) to cover trash and organic matter. The plough bottom is the most important part of the plough. To provide clearance for the cutting edge of the plough, the point is directed about 6-8 mm downward (Vertical Clearance) and about 6-8 mm toward the un-ploughed land (Horizontal Clearance). The mould board plough covers about 0.4 ha a day.

With the adoption of paddy on a large scale the disc harrows have become very popular with the farmers in India. This is because of their ability to cut the straw and incorporate it in the soil as well as for better pulverization. The use of disc harrows has become an absolute necessity at farms where combines harvest the paddy. Bullock-drawn disc harrow is one of the most useful tillage implement. The draft of this implement is almost the same as that of a MB plough or soil stirring plough. The main advantages of the disc harrow include thorough pulverization of soil, cutting and incorporating of stubble and weeds in the soil and higher output as compared to a plough.

Animal-drawn disc harrow (Pandey *et al.*, 1997, Pandey and Ganesan, 2005) has usually six or eight discs fixed on two gangs; each gang has 3 or 4 discs (Fig. 3.2). It consists of a beam, adjusting lever, frame, spool, transport wheels, centre shovel and plane or notched discs. The frame is made of mild steel on which gangs with discs are mounted. It is a single acting double gang type disc harrow suitable for secondary tillage operation. The harrow is provided with an operator's seat and a transport wheel which aids in easy transportation. The operator's seat enables the operator to ride instead of walking, which helps in deeper penetration and reduces drudgery. Also the disc angle can be adjusted easily with the help of the slotted bracket and the depth and width of operation can be controlled. It saves 90 per cent labour and operating time and also results in 80 per cent saving in cost of operation compared to conventional method of ploughing by bullock drawn country plough.

Fig. 3.2: Bullock-drawn disc harrow.

Bullock-drawn harrow-cum-puddler consisting of two gangs of six discs of 45 cm diameter has been developed at Punjab Agricultural University, Ludhiana (Fig. 3.3), Garg and Singh (2002), Anonymous (2012). This can be used for both dry seedbed preparation and puddling for rice transplantation. The discs are spaced at 155 mm with the help of circular drum of 220-mm diameter instead of spools. This leaves only 115 mm of exposed edge of disc to penetrate into the soil, which does not allow excessive sinkage of discs. At maximum depth of 115 mm the disc starts floating and reduces draft requirements. A shovel has been provided in between two gangs to loosen the soil left. Two gangs are mounted on a mild steel frame with help of brackets and wooden bushes. Provision is made in the frame to adjust gang angle up to 25 degrees for increasing depth of cut. A seat has been provided on the frame, which eliminates walking of operator behind harrow and provides weights for better penetration. The machine can cover 1.0-1.5 ha/day.

Animal drawn patela harrow is secondary tillage equipment for clod crushing, stubble or trash collection, levelling and smoothening of land surface before seeding (Singh, 2007, Pandey *et al.*, 1997). It consists of a sal wood plank, trash collection hooks (13 nos.), cogwheel and lever for lifting. It is available in 1.5 and 2.0 metre working widths. Field capacity is 0.3 ha/h and labour requirement 3-4 man-h/ha. Animal drawn Naveen bakhar blade is a V-shaped blade, which is modified version of conventional straight blade of harrow (*bakhar*). The *Naveen Bakhar* blade can be attached to the local bakhar frame easily. It is suitable for primary and secondary tillage operations in black cotton soil. It saves 21 per cent labour and operating time and 17 per cent on cost of operation compared to conventional method of using local 'Bakhar' (blade harrow).

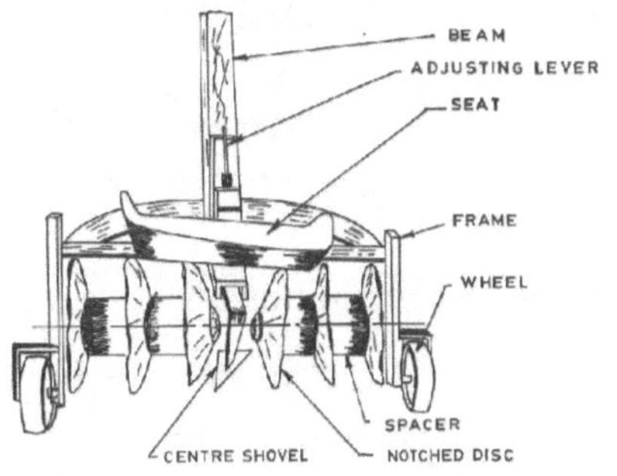

Fig. 3.3: Bullock-drawn disc harrow-cum-puddler.

Spring tooth harrow is a harrow with tough flexible teeth, suitable to work in hard and stony soils (Singh, 2007, Pandey *et al*., 1997, Pandey and Ganesan, 2005). It is fitted with springs, having loops of elliptical shape (Fig. 3.4). It gives springing action in working conditions. It pulverizes soil and helps in killing weeds. It can also be used to loosen previously ploughed soil ahead of grain drill. In this machine, teeth penetrate deeper than spike-tooth harrow and tears out the roots and bring them to surface. This harrow consists of teeth, tooth bars, clamps, frame, clevis, lever and links. The teeth consists of wide, flat, curved, oil-tempered bars of spring steel, one end of which is fastened rigidly to a bar; the other end is pointed to give good penetration. Adjusting angle of teeth by means of levers controls the depth to which teeth will penetrate the soil. The teeth are available with points of various widths and shapes with detachable points for different types of work. Trailing type spring tooth harrows are available in sizes ranging from 2.4 to 11.0 m. The draught of spring-tooth harrow varies from 100-200 kg/m width, depending upon the type of work and soil conditions.

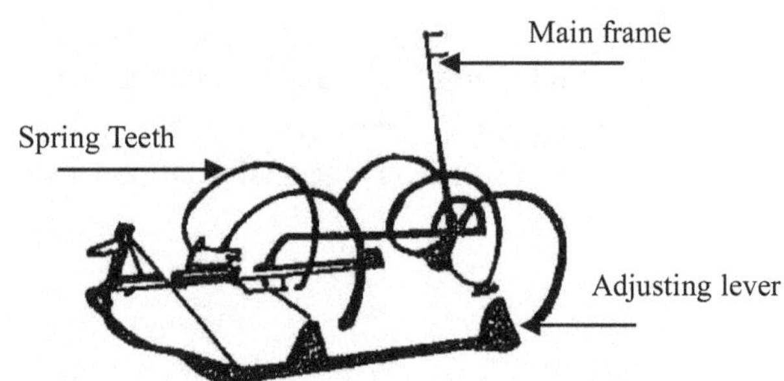

Fig. 3.4: Animal-drawn spring tooth harrow.

Animal drawn blade harrow is also known as Bakhar (Fig. 3.5). It consists of one or more blades attached to beam or frame and used for shallow working in the soil with minimum soil inversion. It is used to prepare seedbed mostly in clay soils. The blade is made of steel and its width varies from 40 to 100 cm. The body of the beam is made of wood.

Animal drawn puddler is a straight/angular/helical/rectangular blade type puddler suitable for puddling wetland and also for cutting and mixing of green manure crops (Fig. 3.6), Pandey *et al.* (1997), Anonymous (2012).

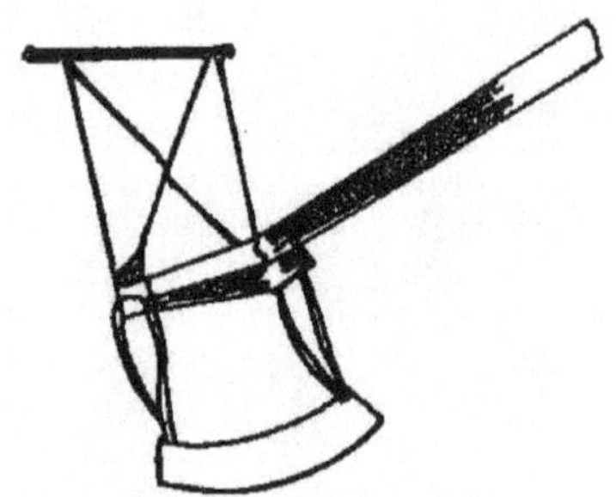

Fig. 3.5: Animal-drawn blade harrow

Width of cut is 700 mm and depth of operation 60-120 mm. It is used for preparation of paddy fields with standing water (5-10 cm depth) after initial ploughing. It breaks up the clods and churns the soil. The main purpose of puddling is to reduce leaching of water and to kill weeds. Puddling facilitates transplanting of paddy seedlings. Puddler consists of puddling units each having four paddles mounted on an axle, frame, beam, metal-cross and handle. Paddles are made of mild steel sheet having thickness of 3.15 mm. While moving, blades (paddles) churn the soil and mix it properly. The weeds are also chopped and mixed with soil for decomposition. Two to three operations are good enough to get desired puddled soil. The animal-drawn puddler can cover 0.4-0.5 ha/day. According to the power used puddler can be classified as hand operated, animal-drawn and tractor-drawn puddler. Animal-drawn puddlers are commonly used in India. It saves 30 per cent labour, 46 per cent operating time and 30 per cent on cost of operation compared to conventional method of using country plough or tractor with cage wheels for puddling.

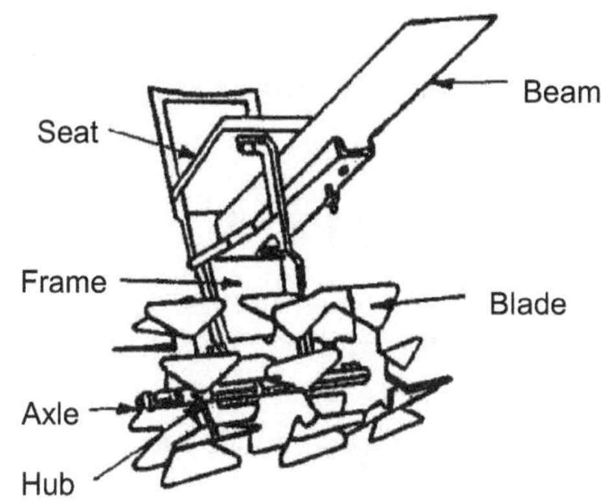

a) Angular Blade Puddler

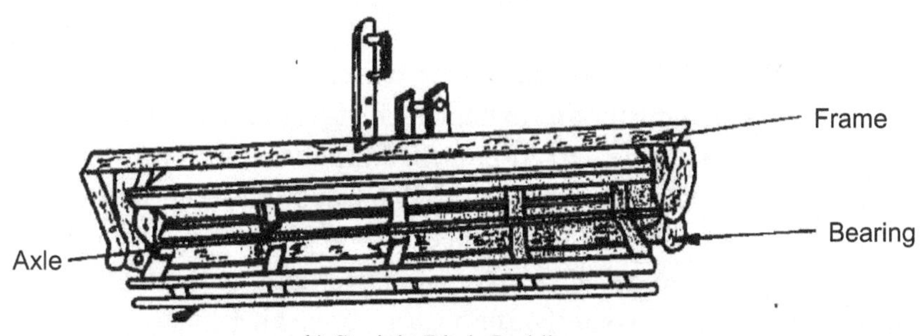

b) Straight Blade Puddler

c) Helical Blade Puddler

d) Rectangular Blade Puddler

Fig. 3.6: Animal-drawn puddlers

Hydrotiller: It is a self propelled, walking type implement with power operated cage wheels suitable for puddling in light and medium soils (Singh, 2007, Pandey *et al.*, 1997). It is operated by 5 hp diesel engine. The field is required to be ploughed once with MB plough before using the machine. Four to five passes are required to achieve desired quality of puddling. It can cover

0.15 ha/h. It saves 76 per cent labour, 87 per cent operating time and 47 per cent cost of operation compared to conventional method of using bullock drawn country plough.

Power Tiller and Tractor Operated

Tractor-drawn implements are used for primary as well as secondary tillage operations. Implements used for primary tillage are mould board plough, disc plough, sub-soil plough and chisel plough. Secondary tillage Implements are disc harrow, spike tooth harrow, spring tooth harrow, rigid tine cultivator and spring tine cultivator.

Different types of ploughs used all over the world may be grouped as indigenous ploughs and soil turning or stirring ploughs (Singh, 2007; Singh and Verma, 2009). Soil stirring ploughs are further classified as mould board ploughs, disc ploughs, sub-surface ploughs, chisel ploughs and other ploughs. There are four types of tractor-drawn implements viz. trailed or pull type, semi-mounted type, mounted type and self-propelled. A pull or trailed type implement is one that is pulled and guided from a single hitch point and is never completely supported by the tractor. A semi-mounted type implement is attached to the tractor through a horizontal hinge axes and is partially supported by tractor during transportation. It is controlled through a 3-point hitch linkage. A mounted type implement is one that is attached through a 3-point hitch linkage and is completely supported by the tractor when in raised position. A self-propelled machine is one in which the propelling power unit is an integral part of the implement.

Mould board plough is normally of two types: fixed or reversible bottom plough and mounted or trailed type ploughs. Fixed bottom ploughs throw the soil in one direction, usually to the right (Fig. 3.7). The plough bottom acts as a 3-sided wedge as it passes through the soil. The main parts of the plough are beam, top link connection, vertical stut, cross shaft, coulter stop, coulter arm, coulter disc, scraper, share point, share wing, mould board and vertical beam (Fig. 3.8). There are different types of bottoms viz. stubble bottom, general purpose bottom, high speed general purpose bottom, slatted general purpose bottom, and breaker or sod bottom (Fig. 3.9). The stubble bottom mould board has an abrupt curvature. It turns the furrow slice quickly and provides maximum granulation. This bottom scours better in sticky soil. It is used for ploughing stubble land and not suitable for high speed ploughing. The general purpose bottom has a longer mould board with less curvature than stubble bottom. It has a fairly long mould board with a gradual twist, the surface being slightly convex. It turns the soil less abruptly and may be used in heavy soil, sod or stubble ground. It may be operated at higher speed than stubble bottom. However, speed should be less when ploughing sod than stubble ground. The high speed general purpose bottom has slightly less curvature in the upper part of the mould board as compared to general purpose bottom. It has a less twisting action and can be operated at a higher speed without excessive throwing of the furrow slice. The slatted general purpose bottom is made of slots placed along the length of the mould board so that there are gaps between the slots. This increases the soil pressure against the remaining portions of the mould board that aids scouring in sticky soils. The breaker or sod bottom mould boards are long with gentle curvature that lifts and overturns the furrow slice. They turn the furrow slice more slowly. It is mainly used in heavy sod or clay soil. It is used in tough soil with full of grasses. This is very useful where total inversion of soil is required.

a) Two bottom MB plough throwing soil towards right.

b) Two bottom MB plough

c) Three bottom MB plough

d) Four bottom MB plough

Fig. 3.7: Tractor operated mould board plough.

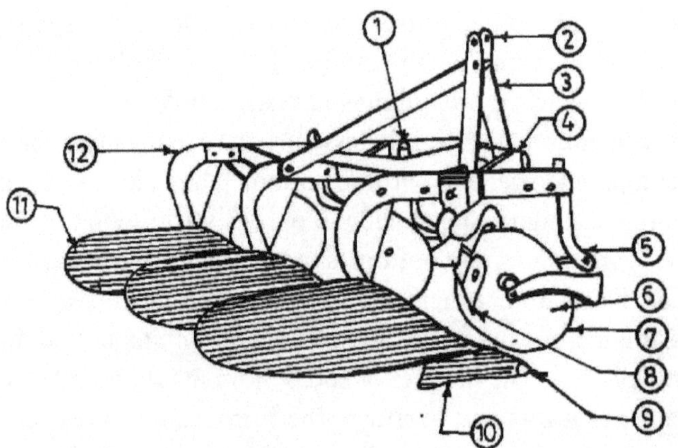

1. Beam
2. Top link connection
3. Vertical stut
4. Cross shaft
5. Coulter stop
6. Coulter arm
7. Coulter disc
8. Scraper
9. Share point
10. Share wing
11. Mould Board
12. Vertical beam

Fig. 3.8: Tractor mounted 3-bottom mould board plough.

There are mainly three accessories used on plough bottoms viz. jointer, coulter and gauge wheel. Jointer is a small irregular piece of metal having a shape similar to ordinary plough bottom. It looks like a miniature plough. Its main function is to cut and turn over a small ribbon like furrow slice directly in front of main plough bottom. It is set to cut 4-5 cm deep. This furrow slice is inverted so that all trash on the top of the soil is completely turned down. Coulter is used to cut furrow slice vertically from the land ahead of bottom and leaves a clear wall. It also cuts trashes covered under the soil. In general, coulter should be set about 5 cm shallower than the depth of ploughing. To obtain a neat furrow wall, the coulter is kept 2 cm outside the landside of the plough. It may be rolling or sliding type. Rolling coulter is a round steel disc suspended on a shank and yoke from the beam. Mostly plain rolling coulters of 30-45 cm diameter are used. The sharpened edge of coulter may be smooth or notched. It can be adjusted up-down and sideways. The up-down adjustment controls depth and sideways width of cut. Sliding coulter does not roll over the ground but slides. The knife may be of different shape and size. Gauge wheel helps in maintaining depth of cut under different soil conditions. It is usually placed in hanging position.

For proper penetration and efficient operation some clearance is provided in the mould board plough and that is called "suction". The suction is of two types viz. vertical and horizontal suction (Fig. 3.10). Vertical clearance is the maximum clearance under the landside and the horizontal surface. Vertical suction is measured at the joining point of share and landside. It helps in maintaining depth of cut. Horizontal suction is the clearance between the landside and the furrow wall. It helps in maintaining width of cut. Mould board

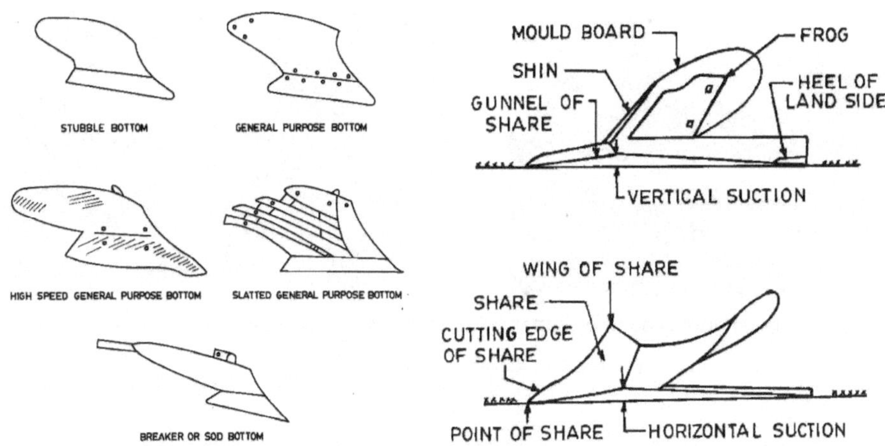

Fig. 3.9: Different types of bottom used on tractor operated mould board plough.

Fig. 3.10: Horizontal and vertical suction on tractor operated mould board plough.

plough needs 3-5 mm vertical clearance (vertical suction) and about 5 mm side clearance (horizontal suction). Vertical clevis is a vertical plate with a number of holes, controls the depth of cut and adjusts the line of pull. Horizontal clevis affects lateral adjustment of plough relative to line of draught.

Reversible bottom ploughs have the bottoms so arranged that the right turning bottoms can be quickly and readily replaced with a set that turns the soil to the left (Fig. 3.11). Thus, when end of the furrow is reached, the plough is raised, turned and returned across the field ploughing into the furrows just made. With reversible ploughs there are no dead furrows, no uneven and sunken spots. Mounted or trailed type ploughs are also called integral ploughs because the tool and power are used as one unit. Hydraulically operated mounted ploughs are easy to handle. The trailing ploughs, which have standard hydraulic cylinder for raising and lowering, are also easy to operate. However, it is more difficult to back trailing equipment than a mounted one. Land wheel is used in raising and lowering the trail-type plough by means of hydraulic or mechanical lift.

Two bottom and three bottom (*Courtesy*: M/s New Holland India Pvt. Ltd.)

Fig. 3.11: Tractor operated hydraulic reversible plough.

Disc Plough: It is plough that cuts, turns and in some cases breaks the furrow slice by separately mounted steel discs (Fig. 3.12). A disc plough is designed with a view to reduce friction by making a rolling plough bottom instead of a sliding plough bottom (Singh, 2007; Singh and Verma, 2009). A disc plough works well in the conditions where mould board plough does not work satisfactorily. The main advantages of a disc plough are: i) works well in too hard and dry soil, ii) works well in sticky soil, iii) more useful for deep ploughing, iv) can be used safely in stony and stumpy soil, and v) works in loose soil also without much clogging. The main disadvantages are: i) not suitable for covering surface trash and weeds, ii) leaves the soil in rough and cloddy conditions, and iii) much heavier as compared to mould board ploughs.

a) Two bottom disc plough being used in field.

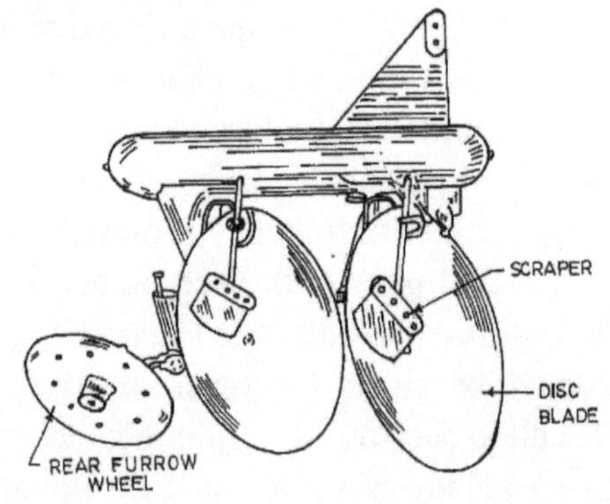

b) Details of two bottom disc plough

Fig. 3.12: Tractor mounted disc plough.

There are two types of disc ploughs viz. standard disc plough (Fig. 3.13) and vertical disc plough (Fig. 3.14). Harrow Plough is also known as vertical disc plough or one way plough. It is a plough that combines the principle of regular disc plough and disc harrow and used for shallow working in the soil. It has a frame, wheel arrangement and depth adjusting devices same as regular disc plough, but the discs are fitted on a single shaft and turn as one unit like a gang of disc harrow. The spacing between discs may be 20-25 cm. Size of the disc varies between 50-65 cm. Disc angle varies from 40-45^0 and tilt angle is zero. Standard disc plough consists of steel disc of 60-90 cm diameter set at a certain angle to the direction of travel. Each disc revolves on a stub axle in a thrust bearing. The angle of disc to the vertical and to the furrow wall is adjustable. In action, the disc cuts the soil, breaks it and pushes it sideways. There is a little inversion of furrow slice as well as little burying of weeds and trashes. Disc plough may be mounted or trailed type. In mounted type disc plough, side thrust is taken by wheel of the tractor. In trailed type, side thrust is taken by furrow wheel of plough. Disc is made of heat treated steel of 5-10 mm thickness. The edge of the disc is well sharpened to cut the soil. The amount of concavity varies with the diameter of the disc and it is approximately 80 mm for 60 cm and 160 mm for 95cm diameter.

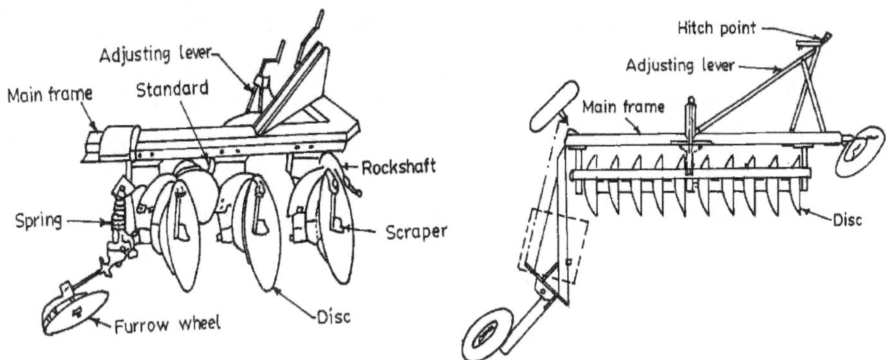

Fig. 3.13: Tractor operated standard disc plough.

Fig. 3.14: Tractor operated vertical disc plough.

Disc angle (Fig. 3.15) is the angle at which the plane of the cutting edge of the disc is inclined to the direction of travel. Usually the disc angle of good plough varies between 42-45 degree. Tilt angle is the angle at which the plane of cutting edge of disc is inclined to a vertical line. Tilt angle varies from 15-25 degree for a good plough. Scraper is a device to remove soil that tends to stick to the surface of a disc. Concavity is the depth measured at the centre of the disc by placing its concave side on a flat surface. Various adjustments can be done on the disc plough to control depth, width and pulverization. By increasing tilt angle, penetration can be improved. However, on standard disc plough penetration is improved by decreasing the tilt angle. Increasing disc angle improves penetration but reduces width of cut. Penetration can be improved by adding weight to the plough. Width of cut on the plough can be adjusted by adjusting angle between the frame and land wheel axle. The plough wheels may be properly adjusted to keep the plough running level.

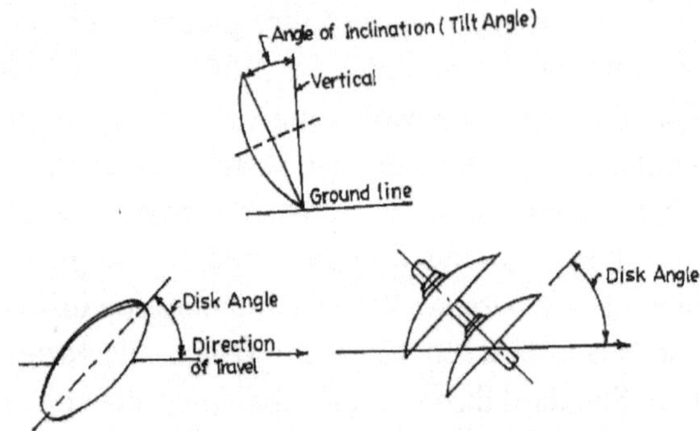

Fig. 3.15: Disc and tilt angle on a tractor operated disc plough.

Disc harrow: A disc harrow is essentially a secondary tillage implement. However, in light to medium soils, heavy-duty disc harrow is also used as a primary tillage implement. A harrow is intended for post ploughing treatment of layers, stubble scuffing to a depth of 8-10 cm and leveling of soil (Singh, 2007; Singh and Verma, 2009). The spacing between the discs on the gang bolt is usually maintained from 15 to 22 cm for light duty harrows and 25 to 30 cm for heavy-duty harrows. Scraper is provided on each disc to prevent soil clogging. It removes the soil that may stick to the concave side of the disc. Better penetration of discs can be obtained by increasing the disc angle, lowering the hitch point, adding more weight on the harrow, using the sharp-edged discs of smaller diameter and lesser concavity, and regulating the optimum speed. Repair of disc harrow includes sharpening of discs, elimination of defects in disc gangs and frame parts. The spherical discs are ground on the convex side to an angle of 50 degree. Discs are made of high-grade heat-treated hardened steel. Tractor-drawn disc harrows have concave discs of size varying from 35 to 70 cm diameter. Concavity of disc affects penetration, inversion and pulverization of soil. All nuts and bolts must be checked daily before transporting the implement to the field. Blunt edges of the discs should be sharpened regularly.

Harrows are used to break the clods, cut weeds, pulverize soil, cover seeds and smoothen the surface (Singh, 2007; Singh and Verma, 2009). There are many types of harrows viz. Disc harrow, spring tooth harrow, spike tooth harrow

(Fig. 3.16), triangular harrow (Fig. 3.17), blade harrow (Fig. 3.18), guntaka harrow and acme or knife harrow. Generally, the disc harrows are of 2 type viz. Single action and Double action (Fig. 3.19). Tandem disc harrow and offset disc harrows are double action type harrows (Fig. 3.20). Disc harrows are also mounted type (Fig. 3.20) and trailed type (Fig. 3.21). The most commonly used disc harrow is offset disc type. Offset disc harrow has two gangs in tandem. It consists of tool bar, front and rear disc gangs, gang angle adjustment, spacing spool and steel discs. It is capable of being offset to either side of the centre line of tractor. It travels to the left or right of the tractor. The line of pull is not in the middle and that is why it is called offset disc harrow. It is generally used for cultivating under the trees. The soil is worked twice in one single operation in opposite direction with this harrow. It also cuts the weeds, which get ploughed with the soil. Two operations with these implements are sufficient for good seedbed preparation in light and medium soils. The average capacity of this implement is about 2.5-3.0 ha/day.

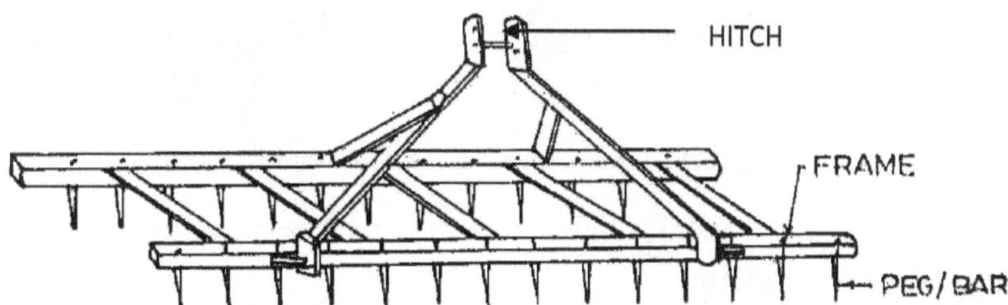

Fig. 3.16: Tractor mounted spike tooth harrow

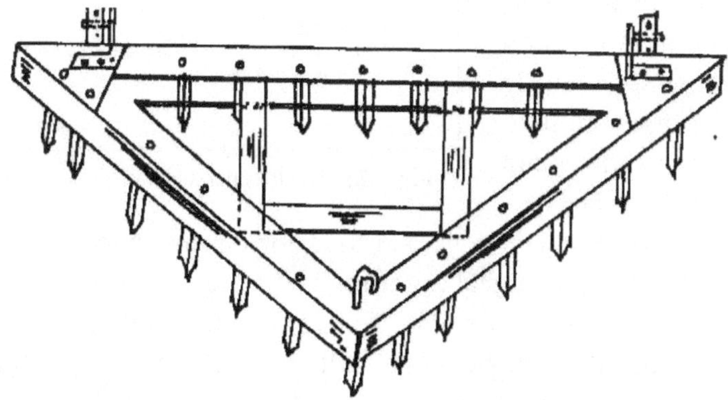

Fig. 3.17: Triangular harrow

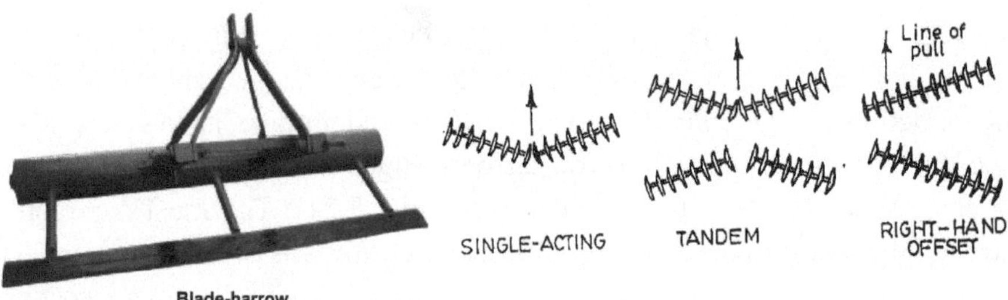

Fig. 3.18: Tractor mounted blade harrow

Fig. 3.19: Different types of disc harrows.

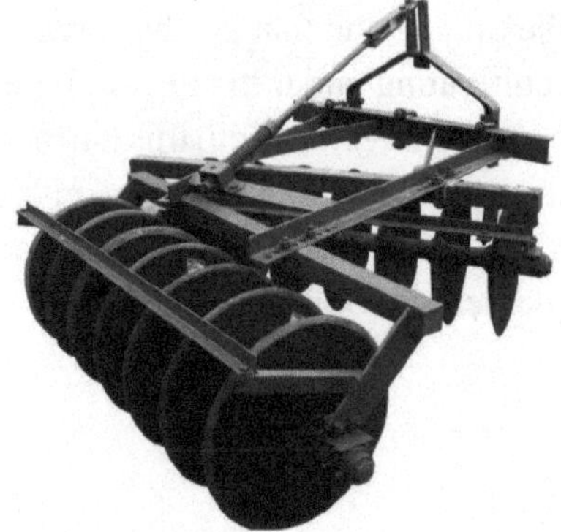

Fig. 3.20: Tractor-operated offset disc harrow

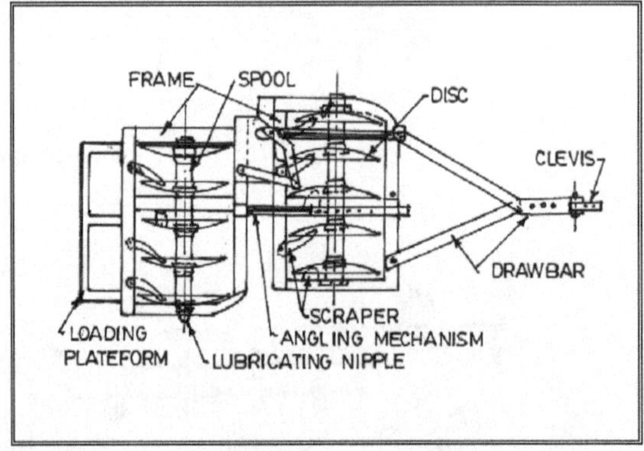

Fig. 3.21: Tractor operated trailed type disc harrows.

Disc is a circular, concave revolving plates used for cutting and inverting the soil. Tractor-drawn disc harrows have concave disc of 35-70 cm diameter. Concavity of disc affects penetration and pulverization of soil. Usually two types of discs are used on disc harrows viz. plain and cut away or notched type disc. Plain discs have plain edge and used for normal work. Notched discs have serrated edge and cut stalk, grasses and vegetative matters better than plain discs. Cut away discs are not effective for pulverization of soil but it is very useful for puddling the field especially for paddy cultivation (Fig. 3.22). Higher penetration on disc harrow can be obtained by increasing disc angle, by adding additional weight, by lowering hitch point, by using sharp edge

Fig. 3.22: Paddy disc harrow

discs of small diameter and lesser concavity and regulating the optimum speed. Bearings of disc harrows must be thoroughly greased at regular interval and blunt edges of discs should be sharpened regularly. Scraper is provided on each disc to prevent soil clogging. It removes the soil that may stick to the concave side of the disc.

Spike tooth harrow is a harrow with peg shaped teeth of diamond cross-section attached to either rectangular or a triangular frame. The triangular frame is made of wood and is pulled by a pair of animals. The rectangular type spike tooth harrow frame is made of steel and is pulled by tractor. The spike tooth harrow is also known as peg-tooth harrow, bar harrow, drag harrow, section harrow or smoothing harrow. Its principal use is to smooth and level the soil directly after ploughing. It is used to break clods, stir the soil, uproot the weeds, level the ground, break the soil crust and cover the seed. It can be used to cultivate corn and cotton and other row crops in early stages of growth. It is of two type rigid and flexible type. Animal-drawn spike tooth harrows are usually of rigid type. The teeth of spike tooth harrows are made of hardened steel. The teeth may be square, triangular or circular in section. It can stir the soil up to the depth of 5 cm.

Cultivator: It is secondary tillage implement consisting of a number of tines attached to a frame that is used for tilling the soil. The cultivator stirs the soil and breaks the clods. The tines fitted on the frame of the cultivator comb the soil deeply in the field. A cultivator performs functions intermediate between those of plough and the harrow (Singh, 2007; Singh and Verma, 2009). Destruction of weeds is the primary function of a cultivator. The important functions performed by a cultivator are to destroy the weeds in the field, to aerate the soil for proper growth of crops, to conserve moisture by preparing mulch on the surface, to sow seeds when provided with sowing attachments and to prevent surface evaporation and encourage rapid infiltration of rainwater into the soil. Tractor operated cultivators are available in 7, 9, 11 and 13 tines. According to the source of power used, cultivators may be classified as hand operated, animal-drawn and tractor-drawn. Cultivators are of two types viz. cultivator with spring loaded tines (Fig. 3.23) and with rigid tines (Fig. 3.24). Depending upon the type of soil and crop grown, shovels of different types are used such as straight shovel, reversible shovel, hoop or spear head shovel, sweep, furrower, and half sweep (Fig. 3.25). Depending upon local conditions, soil and climate, different types of animal-drawn cultivators have been designed and used extensively. Three-tine cultivator with seeding attachment is very common in most part of the country. It is used for secondary tillage operations, inter-culture of row crops and sowing of cereal crops.

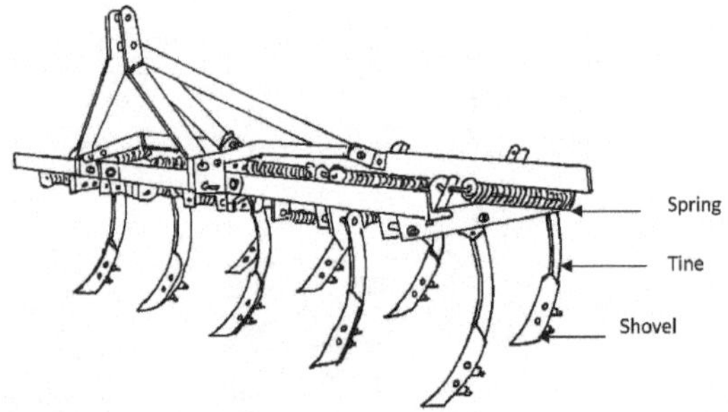

Fig. 3.23: Tractor mounted spring tine cultivator.

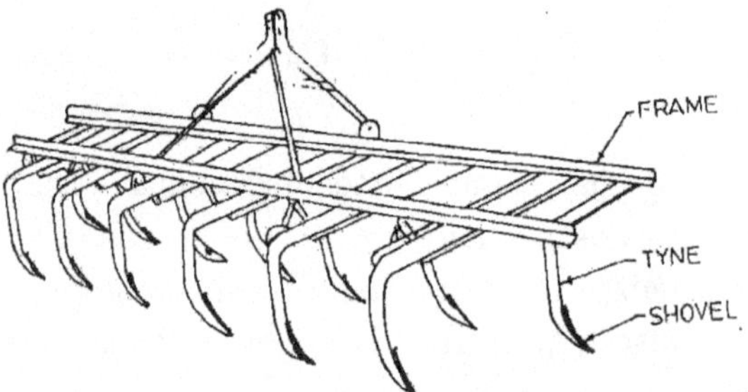

Fig. 3.24: Tractor mounted rigid tine cultivator.

Cultivators having tines loaded with springs so that it swings back when an obstacle is encountered are called spring tine cultivators. Each tine in this cultivator is provided with two heavy coil springs (Fig. 3.23). The springs operate when_shovel point encounters with crop roots or stones during the field operation. On passing over the obstruction, the tines automatically trips and work continuously without interruption. Tines are made of high carbon or spring steel. These types of cultivators are generally used in soils having stones and stumps and are pulled by tractors. Rigid tine cultivators have tines fixed to the frame and tines do not deflect during the field operation (Fig. 3.24). Spacing between tines may be achieved simply by sliding the braces to the desired position or with the help of nuts and bolts.

Duckfoot cultivator: The duck foot cultivator (Pandey *et al.*, 1997) consists of a box type steel rectangular frame, rigid tines and sweeps (Fig. 3.25). The sweeps are triangular in shape similar to foot of duck, hence called duckfoot cultivator. The sweeps are made from old leaf spring steel and fixed to tines with fasteners, which makes them replaceable after being worn out or becoming dull. The tines are made of mild steel flat and forged to shape. It is a tractor-mounted implement and depth of operation controlled by hydraulic system. The sweep cultivator is popular in black cotton soils. This cultivator is mostly used in hard soils for shallow ploughing. It has 5, 7 or 9 sweeps. It can be operated by 25-50 hp tractors. It is used for primary tillage operation, destruction of weeds and retention of soil moisture.

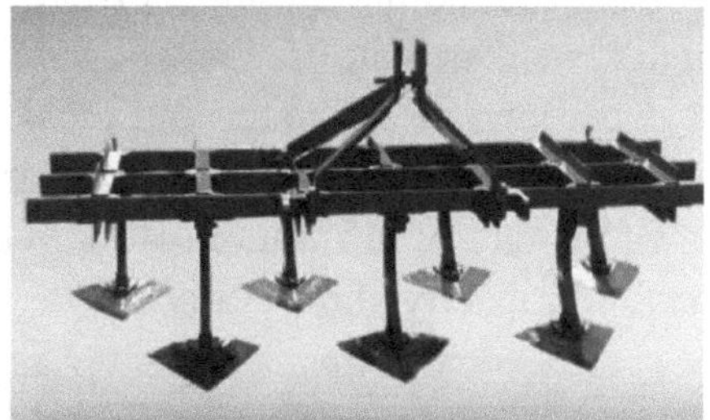

Fig. 3.25: Tractor mounted duckfoot cultivator.

Tractor-mounted pulverizing roller attachment to cultivator: A pulverizing roller attachment for tractor-drawn cultivator has been developed at Punjab Agricultural University, Ludhiana (Fig. 3.26) for breaking the clods and preparing a smooth seedbed (Garg and Singh, 2002; Anonymous, 2012). It consists of an axle on which different numbers of star wheels are mounted depending upon the length of roller. On the star wheels there are six spikes, through each of which pulverizing member passes. The pulverizing members run in a helix from one star wheel to next star wheel. The roller can be attached or detached from the cultivator with the help of a few nuts and bolts. The roller is made of mild steel flates. Tractor-drawn pulverizing roller attachment for cultivator is mounted at the back of the cultivator. The pulverizing roller can prepare a good quality seedbed/puddle in 2-3 operations as compared to at least four operations with the cultivator alone. The field capacity is 0.8 ha/h for dry cultivation and 0.4 – 0.5 ha/h for pudding operation. The savings with this machine in fuel & time are 20-35% and 20-30% respectively. Additionally, due to better puddling, it saves 30-40 percent water requirement for paddy fields. It has been found very useful for wetland and dry land cultivation. It pulverizes the soil thoroughly and makes it good for proper germination. It has also been found quite useful for puddling the field for rice transplanting. It slightly increases the draft and power requirements. But it reduces the number of operations of cultivator and plank. The use of roller with cultivator for puddling also reduces the number of irrigations required for paddy by cutting down the rate of infiltration. However, attaching roller to it does not reduce the field capacity of the cultivator.

a) Pulverizing roller attachment to cultivator in operation under wetland conditions.

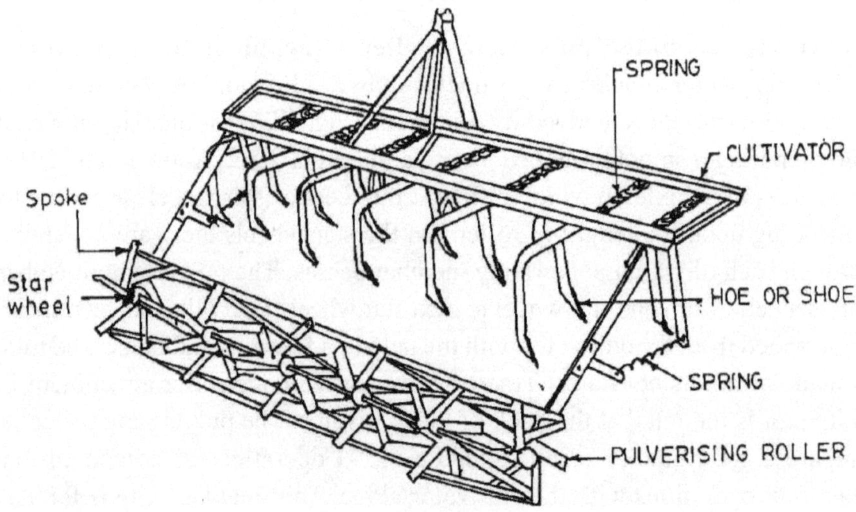

b) Schematic view of tractor mounted cultivator with pulverizing roller attachment.

Fig. 3.26: Tractor operated pulverizing roller attachment to a cultivator.

Courtesy: VST Tillers Tractors Ltd. Bangalore

Fig. 3.27: Power tiller operated rotary attachment

Power tiller with rotary attachment: It is a single axle power unit to till the fields in dry condition and puddle the fields for rice crop (Fig. 3.27). With other attachments it can plough fields, make ridges and interculture in the crops. It can renovate pastures, cut grasses, and level fields, dig potatoes, spray insecticides, pump water and haul up to 1.5-ton load.

Tractor mounted rotary tiller: Tractor operated rotary tiller (also called Rotavator) is suitable for preparing seedbed in a single operation both in dry and wetland conditions (Fig. 3.28). It is also suitable for incorporating straw and green manuring (Singh and Pandey, 2008; Anonymous, 2008, 2008a, 2010, 2010a, 2012). Pulverizing of soil is more uniform and better because impact of revolving blades of rotavator sheares the soil and makes the soil fine. The mean clod size varies between 1.6 - 2.3 mm for rotavator at 210 rpm rotor speed whereas it varies from 4.32 to 5.76 mm for tractor-drawn tiller and harrow. The extent of residue incorporation is above 97 percent for rotavator and only 65-85% when conventional tillage equipment like tiller and harrow is used. The use of a rotavator leads to about 25-40% and 15-25% saving in the fuel consumption for paddy and wheat harvested fields respectively as compared to the conventional tillage implements. It gives highly satisfactory performances while using it under combine harvested wheat and paddy fields as compared to conventional tillage equipment and saves about 40-60% of time.

It consists of a frame, a rotary shaft mounted with blades and power transmission system from the gearbox to the shaft. The blades are of L-type. The rotavator is mounted on three-point linkage system of tractor of 26 kW or higher and it is powered through the tractor PTO. For this operation the torque from the power take-off shaft of the tractor is transmitted to the reduction gear through cardan transmission (cardan shaft). The telescopic shaft and hinges are housed in a protective casing. A friction coupling is provided at the front end of the gear shaft, which protects the tractor transmission from breakage when the cutter drum is overloaded. The gearbox is located at the top of the casing. It has a pair of spur and pair of bevel gears. The gearbox is covered at the back with a removable cover under which two changeable spur gears are located. But this is not so in most of the models currently being operative. The change in turning speed of the cutter drum is then achieved by changing the sprocket in the chain-drive casing. A chain drive is located at the side of the tiller and links the reduction gear to the drum shaft. A tensioning device consisting of a hinged curve strip and a spring is provided to tension the chain. The tubular shaft of the cutter drum rotates between two ball bearings. Flanges are welded to the shaft at regular intervals. The L-shaped cutters three each on the left and right sides are bolted to the flange. The cutters are arranged along a spiral on the shaft. The depth of the working by the tiller depends on the position of the skid or wheel at the left side of the tiller. Fuel consumption varies between 5 to 6.5 l/h with an output of 0.4 to 0.5 ha/h. Break down in good rotavator is very less. However, they require regular maintenance. Rotavator blades are replaced after 250-300 hours of operation. It consists of a power driven shaft on which knives or tines are mounted to cut the soil and trash. Rotor has got several types of tines fitted on the shaft having a speed of 200-300 rpm.

According to power used, rotavators are classified as animal-drawn, engine operated and tractor-drawn rotavators. One or two operations of this implement are sufficient for good pulverization of soil depending upon soil and crop conditions. It is not meant for sandy soil. The depth of penetration can be adjusted up to 12.5 cm. The suitable protective cover is provided at the rear to prevent undue scattering of soil. Bullock-drawn engine operated rotary tiller is quite useful for timely preparation of seedbed particularly in rice-wheat rotation. Power tiller operated rotary tillers are also quite useful for hilly areas and small land holdings and wetland cultivation. It saves water, time, energy, labour and money. It can be used for green manuring of dhaincha, moong and other crops in wet conditions followed by heavy planker in standing water for puddling, incorporates residues of paddy, sugarcane, cotton in the soil in single operation, better quality of work in less number of operations as compared to traditional practice both in dry and wetland condition, level of field not disturbed, and used throughout the year.

a) Tractor-operated rotavator incorporating weeds.

b) Tractor operated rotavator under operation in dry and wet conditions.

Fig. 3.28: Tractor operated rotavator (Rotary tiller)

Chisel Plough: It is an important implement used to cut through hard soils by narrow tines. It is used before using the regular plough (Singh, 2007). It is useful for breaking up hard layers of soil just below the regular ploughing depth also called as hardpan or plough sole. As stirring breaks the soil, it is not inverted and pulverised to the extent mouldboard and disc ploughs do. The chiselling and stirring operation does not throw enough soil to cover trash completely and hence, often used for stubble-mulch or sub-surface tillage practices. Special plough shovels, lister shovels, sweep and knife assemblies and seeding attachments make it possible to do many jobs with the chisel-type plough.

Chisel plough is a tool with rigid curved or straight shank with relatively narrow shovel points. It may be termed a heavy-duty deep cultivator. The standard or shank is made of nickel-alloy heat-treated spring steel that is given a long and gradual curve flat-wise to permit a slight spring action. The standards are arranged on heavy frames in two or three staggered rows to permit trash to pass between them without choking. Most chisel ploughs are provided with coil cushion springs in conjunction with clamps to allow the tool to swing back. Pneumatic-tired wheels control the depth as well as serve for transportation. Hydraulic lifts are provided for all sizes of chisel ploughs.

Auger Plough: This implement is a combination of the mould board plough and vertically rotating auger (Fig. 3.29), Singh (2007). Only short mould boards are used to provide sufficient space for vertically rotating blades. The furrow slice cut by share slide over the mould board and strikes against the auger. Because of high speed of rotating blades, the soil is pulverised to a very fine size. It buries the weeds and useful for green manuring also. It can also be used for puddling the field. The average field capacity of a 3-bottom auger plough operated by 60-hp tractor varies from 2-2.5 ha/day.

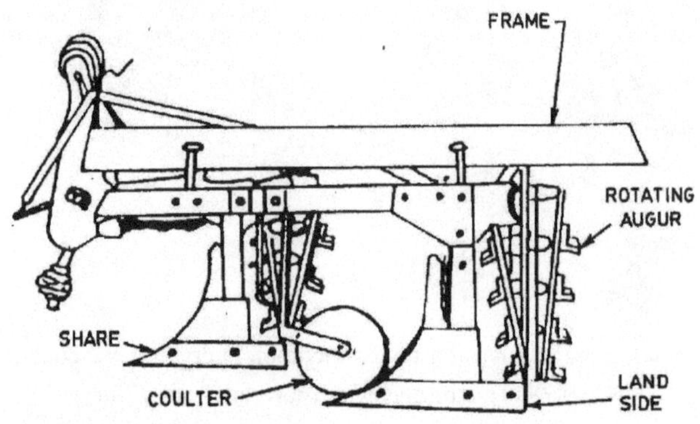

Fig. 3.29: Tractor operated rotating auger plough

Sub soiler: It is like a chisel plough with a single tine that may penetrate up to 100cm depth or more and hence, made heavier than chisel plough (Fig. 3.30), Singh (2007); Pandey *et al.* (1997). Tractor of 60-85 hp is required for pulling single tine sub soiler. One standard is used for deeper depth, but two or three can be used for shallower operations. It is used to plough the soil for orchards, vineyards and forest plantation. It can also be used to break the hard pan formed due to long cultivation or under-plough layer up to the depth of 60-80 cm. It consists of a plough bottom, share and mould board surface. Ahead of plough bottom, a special chisel nose is installed that helps the plough to penetrate deeper into the soil and protects the share point against damage. Depth adjustment holes are provided on the sub soiler. There is a screw type mechanism for adjusting plough depth and horizontal levelling of plough. Sub soiler is available in both trailing and mounted type. It can cover about 1.2-1.5 ha/day.

Sub-surface plough: It is just like a chisel plough capable of penetrating to a depth of 50-90 cm. Tractors of high horse power are used to operate this plough. It is heavier than chisel plough.

Bund Former: It is used for making bunds or ridges by collecting soil (Singh,

Fig. 3.30: Tractor operated sub soiler.

2007; Pandey *et al.*, 1997). Bunds are required to hold water in the soil, thereby conserving moisture and prevent runoff. The size of bund former is determined by measuring the maximum horizontal distance between the two rear ends of forming boards. It is operated by both animal and tractor. It consists of forming board, beam and handle (Fig. 3.31). Forming board is made of mild steel of thickness 1.6 mm for light soil and 2.0 mm for medium and heavy soils.

Ridger: It is an important implement used to form ridges required for sowing

(a) Animal-drawn bund former b) Tractor operated bund former

Fig. 3.31: Bund former

row crops seeds and plants in well tilled soil (Singh, 2007; Pandey et al., 1997). It is also used for forming field furrows or channels, earthing up and similar other operations. Ridgers are also known as ridging plough or double mould board plough. Both animal-drawn and tractor operated ridgers are available in varying sizes (Fig. 3.32). The width of the ridger can be changed by adjusting the ridger blades.

Land rollers and pulverizers: They are used for further pulverization of

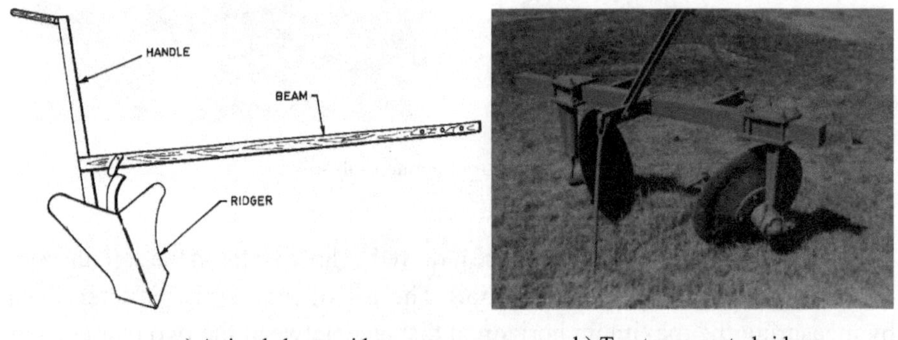

a) Animal-drawn ridger. b) Tractor operated ridger

Fig. 3.32: Ridgers

seedbed especially in heavy and medium soil where clod formation is there (Singh, 2007; Pandey et al., 1997). They are of two type viz. surface and sub-surface packers. There are different kinds of surface packers named according to the shape of roller surface such as V-shaped roller-pulverizer, combination T-shaped and sprocket-wheel pulverizer and flexible sprocket-wheel pulverizer. The sub-surface packers consist of V-shaped packer and crowfoot packer. It rolls, pulverizes, packs, levels, cultivates and mulches the soil in one operation.

The front and rear roller gangs are spaced apart and spring-harrow teeth are mounted between the sections to break the soil and bring clods to the surface so that rear roller gang can crush them.

Normally two types of rollers are used; one flat roller and another ring roller. Flat rollers are used for levelling grassland. It consists of two or more cylindrical cast iron sections. The longer the roller more is the number of sections required for easy turning at the corners. Ring roller consists of a series of cast iron rings mounted on a shaft. Ring rollers are better than flat rollers as the former has better clod crushing effects. Ring roller follows the ground contour better as compared to flat roller because each ring is separately and loosely mounted.

Cotton stalk uprooter (Cotton Stalk Puller): Cotton is one of the important crops grown in the country. After the picking is over, the stalk of cotton need to be removed from the field to prepare the seedbed for next crop. The usual practices followed are either to cut the stalk with an axe and remove the stumps with a digging tool or to pull the stalks manually by hand after irrigating the field. Considerable time and labour is thus involved in this operation. Apart from this manual pulling results in injury to the hand. A cotton puller can pull the stalk in the irrigated field but there is not much labour saving. A manual stalk puller can be used to pull individual stalk from the field, but is very laborious and tedious operation. Tractor operated cotton uprooter has also been developed and it can uproot the cotton stalks.

Gupta (2006) developed a cotton stalk puller that consists of main frame, pulling wheels and shafts, compression spring unit, and transmission system (Fig. 3.33). Main frame of the cotton stalk puller was made from a hollow MS square section of 65 x 65 mm. Two pieces, each of 2000 mm length are kept parallel and ends of both are welded to MS flat of 400 x 65 x 8 mm to get the rectangular shape of the frame. Under the rear section, one more piece of the same length (2000 mm) is attached leaving a gap between upper and lower section so that tynes of the pulling wheels could be mounted on the lower section using U-clamps. Tyne for live wheels (driver wheel) is prepared from a channel section of 150 x 75 mm. Tyne for dead wheel (driven wheel) is made from a MS square section of 65 x 65 mm. Upper end of 500 mm section is fastened to a special type of clamp so that it could be moved freely on both the sides (left and right) to adjust the gap between a pair of pulling wheels. At the lower end of the section a piece of MS shaft of 350 mm is welded to mount the pulling wheel on the end of the shaft. The wheel could rotate freely on this shaft. Two such tynes are mounted in the middle of the frame with the help of U-clamps. The spacing of these wheels on the frame could be adjusted by sliding these tynes depending upon the crop row spacing.

Pulling wheel is the most important part of the machine as it has to pull the

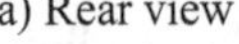

a) Rear view

b) Side view

Fig. 3.33: Cotton Stalk Puller

plants from the field. Rubber coating on the periphery of the wheel is provided by the process of heat treatment and pressure application. Thus, a hardness of 80 N/mm^2 could be achieved. After cooling, the wheel is tapered at one side of the periphery by 17 degrees to get V-shape groove of about 40 mm when both the wheels (live and dead) are kept in contact. Shaft for live wheels (drive wheels) is made from a cold rolled MS bar of 300 mm piece of 35 mm diameter. Shaft for dead wheel (driven wheel) is made from a cold rolled MS bar of 350 mm length and 35 mm diameter.

Fig. 3.34: Tractor operated blade type cotton stalk uprooter.

The ground wheel unit consists of standard, clamp and wheel. The compression spring unit consists of the shaft, hollow Pipe and spring. The transmission system of the machine consists of the Main shaft, Sub-main shaft, Spur gear and Chains and sprockets. The power from tractor PTO is received at main shaft with the help of a universal joint. From main shaft, it is transferred to sub-main shaft through chains and sprockets of varying sizes and ultimately to the pulling wheels through sprockets of the sub-main shaft and extension shaft. Plant pulling efficiency of the machine is largely affected by pull angle and peripheral speed. With the increase in pull angle, plant pulling efficiency increases and was observed highest (61.0%) at 30 degree pull angle and 4.19 m/s peripheral speed of pulling wheels for 2 km/h forward speed of operation.

In an another attempt blade type cotton stalk uprooter has been developed at Punjab Agricultural University, Ludhiana (Fig. 3.34), Singh (2007). This consists of 1-1.2 m width blade mounted on a rigid frame. This is operated by tractor PTO and depth can be controlled by 3-point linkage. It digs the cotton stalks and lefts on the surface which is collected manually for fuel purpose.

References

Anonymous. 2008. Research Highlight. AICRP on Farm Implements and Machinery, CIAE Bhopal. Technical Bulletin No.: CIAE/2008/141.

Anonymous. 2008a. Success Stories.AICRP on Farm Implements and Machinery, CIAE Bhopal. Extension Bulletin No.: CIAE/FIM/2008/80.

Anonymous. 2010. Research Highlight. AICRP on Farm Implements and Machinery, CIAE Bhopal. Technical Bulletin No.: CIAE/2010/151.

Anonymous. 2012. Directory of Successful Farm Machinery in SAARC Countries. SAARC Agriculture Centre. BARC Complex, Farmgate, Dhaka – 1215 (Bangladesh).

Garg I K; Singh Surendra. 2002. Farm equipment for Punjab agriculture. Department of Farm Power & Machinery, Punjab Agricultural University, Ludhiana.

Gupta Ram Autar. 2006. Design, Development and Evaluation of a Tractor Operated Cotton Stalk Puller. Unpublished Ph.D. thesis. PAU Ludhiana.

Pandey M M; Ganesan S. 2005. Farm Mechanisation Package for Dryland Agriculture, Technical Bulletin No: CIAE/FIM/2005/117. Central Institute of Agricultural Engineering, Bhopal.

Pandey M M; Majumdar K L; Singh Gyanendra; Singh Gajendra. 1997. Farm Machinery Research Digest. Technical Bulletin No. CIAE/97/69, Central Institute of Agricultural Engineering, Bhopal, 328 p.

Singh Surendra. 2007. Farm Machinery – Principles and Applications. Directorate of Information & Publication of Agriculture, Indian Council of Agricultural Research, Krishi Anusandhan Bhawan-I, Pusa Campus, New Delhi.

Singh Surendra; Pandey M M. 2008. X Plan Achievements (2002-2007). AICRP on Farm Implements and Machinery, CIAE Bhopal. Technical Bulletin No.: CIAE/2008/137.

Singh Surendra; Verma S R. 2009. Farm Machinery Maintenance & Management. Directorate of Information & Publication of Agriculture, Indian Council of Agricultural Research, Krishi Anusandhan Bhawan-I, Pusa Campus, New Delhi.

4 Seeding, Planting and Transplanting Machinery

The seed drills and planters developed and manufactured in advanced countries are designed to meet the requirements of farmers for the different crops and agro-climatic conditions. These machines are capable of operating at high working speed, cover the large area per unit time and provide the desired plant emergence and stand in the fields. The machines are available to handle the seeds of cereals, pulses, oil seeds and other crops widely grown in the developed countries. In developing countries even though these machines can be made available but they are beyond the reach of small farmers. The level of mechanization of sowing and planting is at a very low level. The traditional method of sowing behind the country plough still continues and covers the large tracts particularly in small and marginal fields and hill agriculture. When animals are used as draft power, the timeliness of field operations is identified as major constraint for raising the crops. Further when crops are raised using traditional method the non uniform or wide variation is noticed between the rows and along the planted rows of crop due to gaps and bunching effect on seeds. This, however, can be overcome by thinning or dibbling method but it would involve more work for the farmers. The improved sowing and planting machines are must for the developing countries due to four reasons: to increase productivity level of land and crop intensity, to overcome timeliness constraints due to climatic conditions, to overcome constraints due to soil factors, and to meet the specific crop or seed needs. Number of studies have shown that use of these machines have helped in increasing the crop yield levels by 10-15% over conventional methods due to timeliness of operation, better plant establishment and efficient application of inputs (Singh, 2007). Traditional methods of seeding and planting include broadcasting manually, opening furrows by a country plough and dropping seeds by hand, known as *'Kera'*, and dropping seeds in the furrow through a bamboo/metal funnel attached to a country plough *(Pora)*. For sowing in small areas dibbling i.e., making holes or slits by a stick or tool and dropping seeds by hand, is practiced. Multi-row traditional seeding devices with manual metering of seeds are quite popular with experienced farmers.

The basic objective of sowing operation is to place the seed and fertilizer in rows at desired depth and seed to seed spacing, cover the seeds with soil and provide proper compaction over the seed. The recommended row-to-row spacing, seed rate, seed to seed spacing and depth of seed placement vary from crop to crop and for different agro-climatic conditions to achieve optimum yields. The functions of seed-drills and planters are: (i) to meter seeds of different sizes and shapes; (ii) place the seed in the acceptable pattern of distribution in the field; (iii) place the seed accurately and uniformly at the desired depth in the soil; and (iv) cover the seeds and compact the soil around it to enhance germination and emergence. Depending upon climatic and soil conditions, seeds are sown in well-prepared and levelled fields, on ridges, in furrows or on beds. Seeding in furrows is done in arid regions to conserve soil moisture and improve plant growth. Bed planting helps in conserving soil moisture, avoids soil compaction and promotes plant growth.

Seeding and planting methods: There are various methods for placing seeds in the soil. These are:

Row seeding: This method places either seeds, bulbs or tubers in soil at desired depth, if rows are wide to permit inter cultural operations, it is known as crop row planting/drilling.

Broadcasting: The seeds are scattered over the surface of soil by hand or machine. The seed distribution is random over the area.

Hill seeding or planting: Seeds are dropped along rows at equal spacing in clusters.

Check row planting: Seeds are placed in rows in two directions perpendicular to each other thus facilitating inter-cultural operations in both directions.

Machine factors which affect seed germination and emergence: The machine factors that affect the seed germination, emergence and plant stand are seed damage during metering, non-uniformity in placement of seed at proper depth, distribution of seed along the rows, displacement of seed from the row, loose soil getting under the seed, degree of soil compaction, uniformity of soil covers, and uniformity of fertilizer placement in furrow.

Selection of seed drills/ planters: The selection of seed drills/ planters is based on basic information about the crops and their characteristics; soils and climatic conditions during sowing seasons; agronomic requirements of crops and yield levels; source of power available; labour requirements for seeding; socio-economic conditions of farmers; size of holding; availability of spares and repair facilities; ease of operation, calibration and maintenance; safety and operator's comfort; expected level of cost of machine and cost of machine operation; and net benefit expected at farmer's level.

Components of seed drills and planters: The major components of seed drills or planters are described as seed hopper or boxes, metering device, seed tubes, furrow openers, covering devices, compaction devices, power transmission unit, main frame, hitch and controls. The machines are mostly classified according to source of power used to operate it such as manually operated, animal drawn (single and pair operated), power tiller, tractor or self-propelled units. Most of the machines are operated by mechanical power from tractor or engine or driven by ground wheel when pulled by the power source. The recent trend is to operate the seed metering and seed conveying devices by small electrical motor or the pumping devices like blowers or by hydraulic motors. The controls are mostly of mechanical type. Presently electronic devices are increasingly used to achieve level of efficiency and performance. These designs indicate the advances made and level of technology attained in the country.

Seed metering devices: Metering mechanism is the heart of seeding machine and its function is to distribute seeds uniformly at the desired application rates. In planters it also controls seed spacing in a row (Singh, 2007; Singh and Verma, 2009). Seed metering device is the major component of a seed drill. The machine performance is greatly affected by this unit. The seed rates of crops vary from very low to very high rates i.e. from 2 to 3 kg/ha to 100 to 120 kg/ha. The fertilizer rates also vary to a great degree say from 50 to 350 kg/ha or more. To achieve uniform seed rates therefore involves the precision with which the device has been manufactured. In case of multi-row devices it is of utmost important to avoid inter-row variation. A seed drill or planter may be required to drop the seeds at rates varying across wide range. Common type of metering devices used on seed drills and planters are: (i) adjustable orifice with agitator; (ii) fluted roller (standard); (iii) fluted roller (small flutes); (iv) vertical rotor/roller with cells; (v) plate with cells; and (vi) cup feed (Figs. 4.1 and 4.2). Adjustable orifice with agitator metering device regulates seed flow by changing the size of opening provided at hopper bottom. An agitator fixed above seed opening helps in continuous flow of seeds. Many designs of conventional animal and tractor operated machines have adopted this mechanism on account of its simplicity and low cost. Fluted roller (standard) is made of axial or helical flutes machined or cast on an aluminium, cast iron or plastic roller. Rotation of fluted roller in housing, filled with seeds, causes the seeds to flow out from roller housing in a continuous stream. Seed rate is controlled by changing exposed length of fluted roller. Fluted roller with small flutes is used for sowing of small seeds like rapeseed-mustard and sesamum. These can be fitted by replacing the standard fluted rollers on the seed-cum-fertilizer drill. Low seed rates of 3 to 5 kg/ha can be achieved with this metering roller with an accuracy of $\pm$ 10 per cent. Vertical rotors with cells are suitable for metering individual or hill of seeds. The rotor with grooves or cells on its periphery is fixed in the hopper. The size and number of cells on the rotor are according to the size of seed and desired seed rate. A cut-off device is provided above the rotor for regulating the flow of seed to cells. Horizontal, inclined or vertical plate with cell type metering mechanism picks and drops individual seed or a hill of seeds depending on design of cell on the plate. Spacing between seeds/hills is controlled by drive ratio and number of cells on plate. Separate plates are required for sowing different crops. Seed picking cups or spoons are provided on periphery of a vertical plate. When the plate rotates, cups pick seeds from seed hopper and drop them in seed funnel. Size of cups depends on size and number of seeds per hill. This type of metering is used for seeds, which are easily damaged by mechanical devices.

Furrow openers: The design of furrow openers of seed drills varies to suit the soil conditions of particular region. Most of the seed-cum-fertilizer drills are provided with pointed tool to form a narrow slit in the soil for seed deposition (Fig. 4.3). Double end pointed shovel type furrow openers of 100 to 200 mm size are used on seed drills in light to medium soils for medium to deep placement of seeds. Pointed bar type (diamond shaped) furrow openers are used for forming narrow slit under heavy soils for placement of seeds at medium depths. Shoe type openers are used in black soil regions. Seeds are dropped through a tube connected to boot at rear of opener for placement at shallow to medium depths. Runner or sword type openers are used on planters for shallow sowing. Soil over seed flows back in furrow during operation and seeds are covered with a levelling bar, chain or by operating a wooden plank behind the drill.

Seed tubes: The seed tubes towards the furrow openers carry the seeds after they are released from the metering devices. The flexible plastic tubes are widely used (Fig. 4.4). The telescopic high-density plastic tubes with proper connections at the metering device and furrow openers are to be provided for better performance. The seed distribution in furrow is affected by the factors: variation of seed release position, trajectory of seed, displacement caused by impact of seed in the furrow, displacement due to soil backflow, and displacement due to soil covering devices. Ideal condition is to release the seed from the seed tube in the furrow at the relatively zero speed with no time gap. However the release of seed is tried to be kept at low level within the limits to achieve good performance. The displacement of seed in furrow and displacement due to soil is caused by furrow openers and covering devices.

Covering and compacting devices: The commonly used covering device is the levelling boarder chain. The covering devices should cover the seed with soil without causing seed displacement in the furrow or allowing the top dry soil to drop below seed. The various types of seed covering and compacting types of wheels are shown in Figs. 4.5 and 4.6. The compacting devices are used to directly compact the soil over the seed or it is a device to compact the soil on both side along the row; or compacts the soil from the sides to form a slight ridge with top surface open. This is to avoid soil crust formation and seedling to emerge easily. The selection of compacting device is based upon the soil conditions encountered in the field or region.

Calibration of seed drills

Calibrate the drill for accurate application rates as per the following procedures:

1) Level the machine with proper support and raise the ground wheel. Check the ground wheel for smooth rotation.
2) Fill rear compartment of hopper with clean seed and the front with fertilizer.
3) Set the feed plate gap of seed fluted rollers according to seed size and that of fertilizer on the top hole, in case of aluminum fluted rollers, metering units.
4) Take out the delivery tubes from the opener boots and place polythene bags to collect seed and fertilizer.

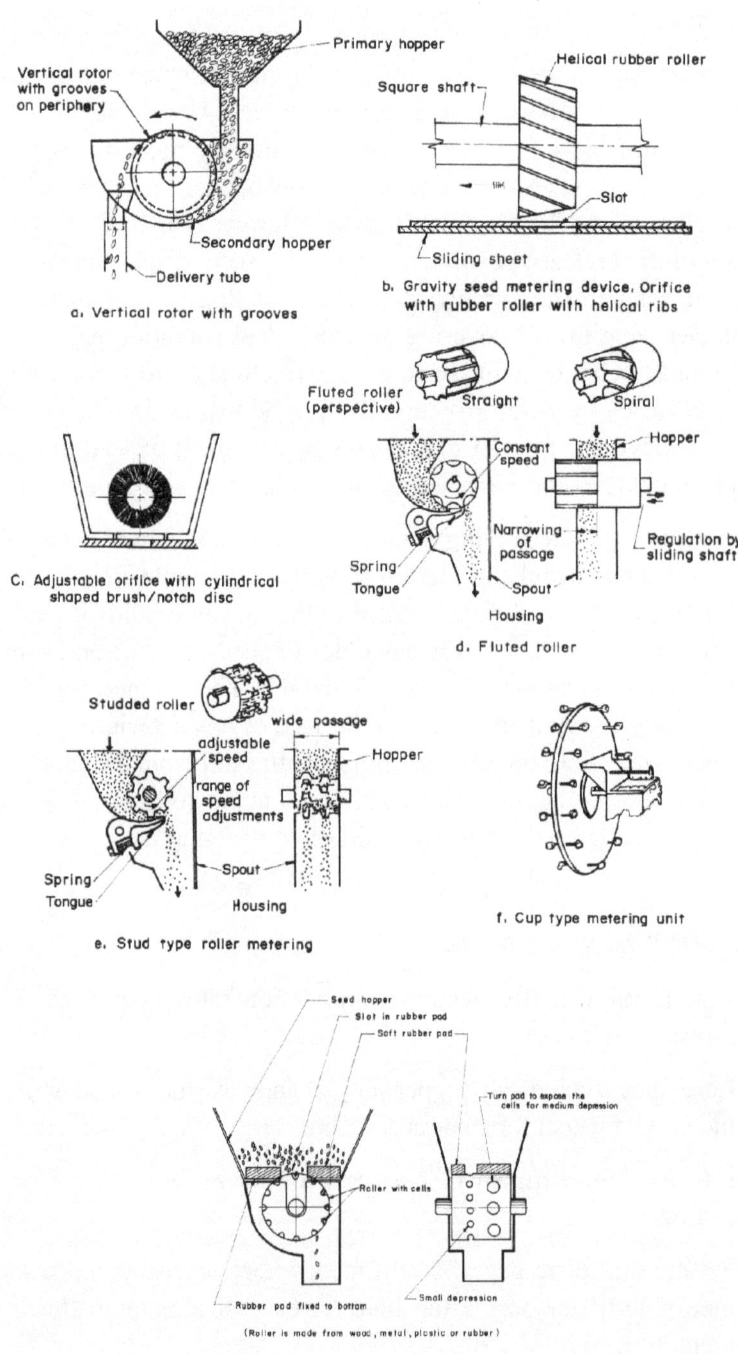

Fig. 4.1: Different types of seed metering devices used in seed drills/planters.

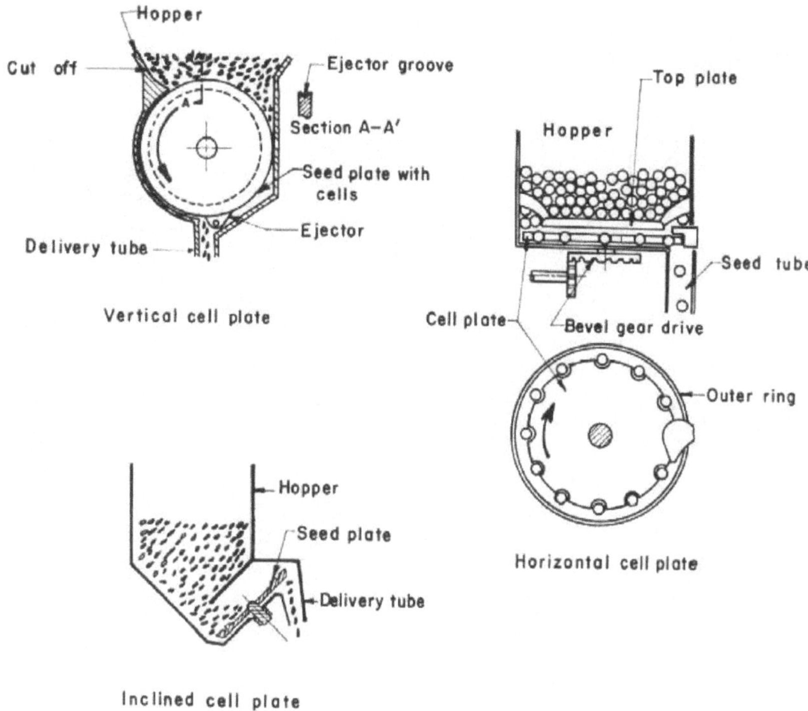

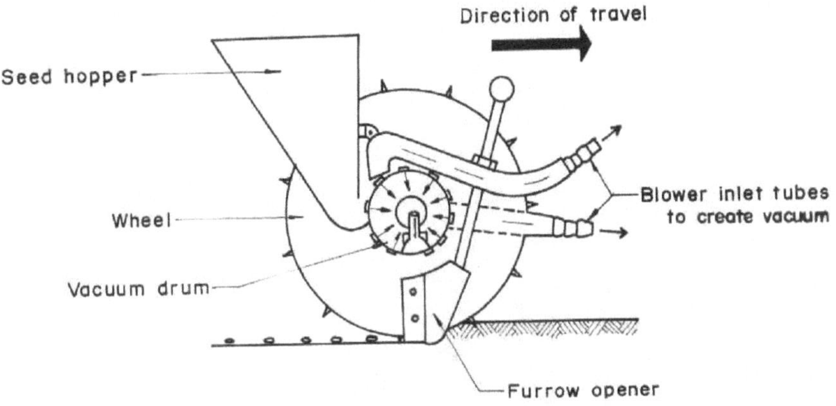

Vacuum-type single seed planter.

Fig. 4.2: Metering system used in planters for dropping single seed.

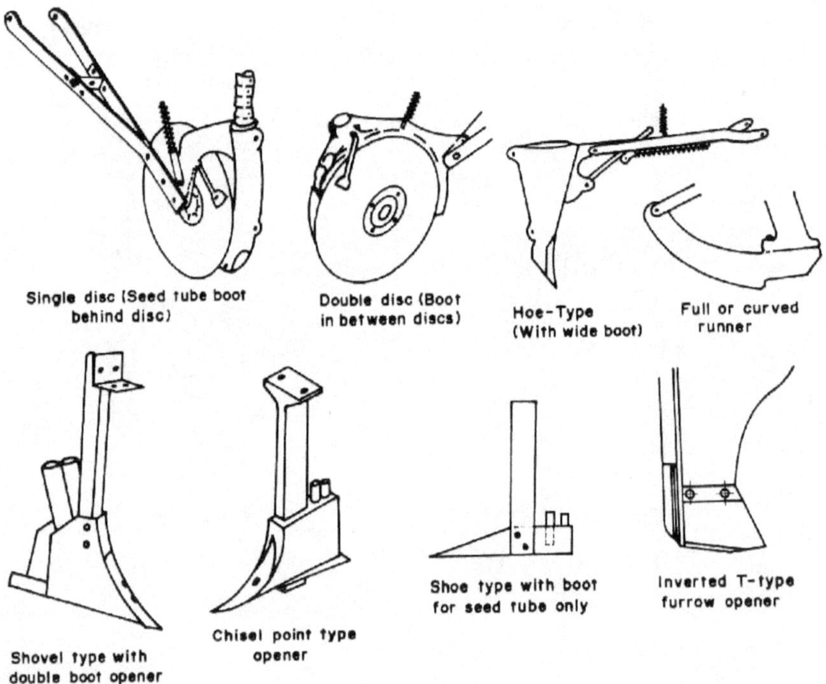

Fig. 4.3: Different types of furrow openers used in seed-cum-fertilizer drill and planter.

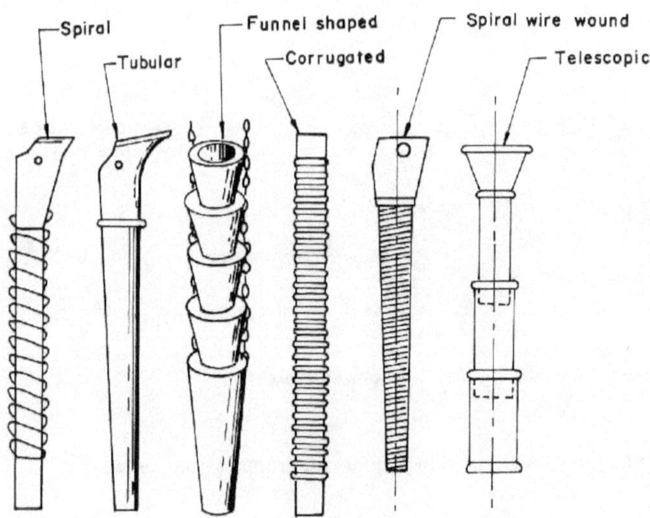

Fig. 4.4: Seed tubes used in seed-cum-fertilizer drill and planter.

5) Turn the ground wheel in forward motion to few rotations first. Remove the seed and fertilizer, so dropped in the bags and again place the bags below the tubes. Now turn the ground wheel to 50 revolutions.

6) Weigh the quantity of seed and fertilizer dropped in 50 revolutions in all the bags and calculate the average amount of seed and fertilizer delivered per opener.

7) Check the values obtained in step 6, with the calculated values obtained by the formulae

$$\text{Seed/fertilizer delivered in 50 revolutions} = \frac{DNSR}{100} n, g$$

Where,

S = seed or fertilizer rate in kg/ha,

n = no. of openers,

R = row spacing in cm,

D = diameter of the ground wheel in cm, and

N = no. of revolution of ground wheel.

8) If the observed values of step 6 are same as that calculated in step 7, the machine is calibrated to sow desired quantities, otherwise proceed to step 9 or 10.

9) If the observed values of step 6 are less than that calculated in step 7 increase the fluted roller exposure by turning the regulator knobs clockwise and repeat step 5 to 8.

10) If the observed values of step 6 are more than that calculated in step 7, reduce the fluted roller exposure by turning the regulator knob anticlockwise and repeat step 5 to 8.

Method of field operations: Align the machine in the field in the direction of operation at one end and lower the machine by raising the transport wheels. Fill clean seed and lump free, dry granular fertilizer in the rear and front compartments, respectively. Set the desired depth of sowing by raising the depth- cum -transport wheels up above the level of opener tips to the extent of the depth of sowing. Run the machine for a short distance and check the depth of sowing. Correct the position of depth wheels to set desired depth of sowing.

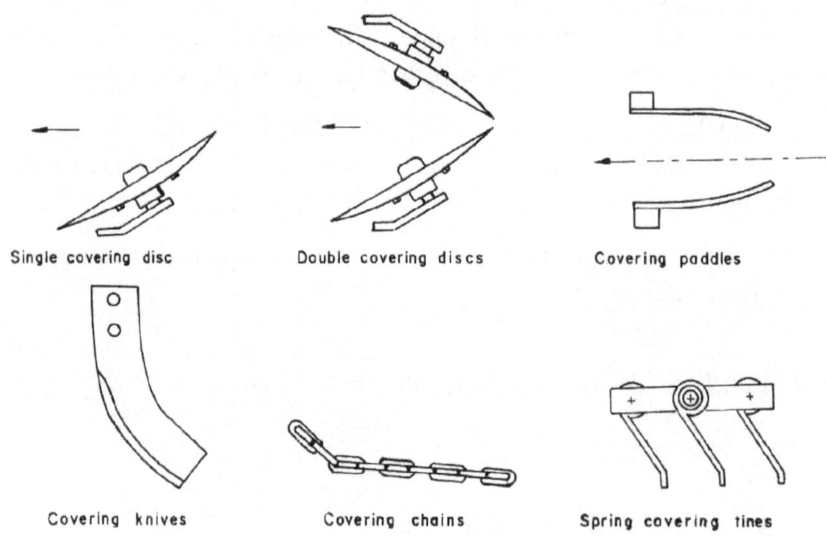

Fig. 4.5: Seed covering devices used in seed drills

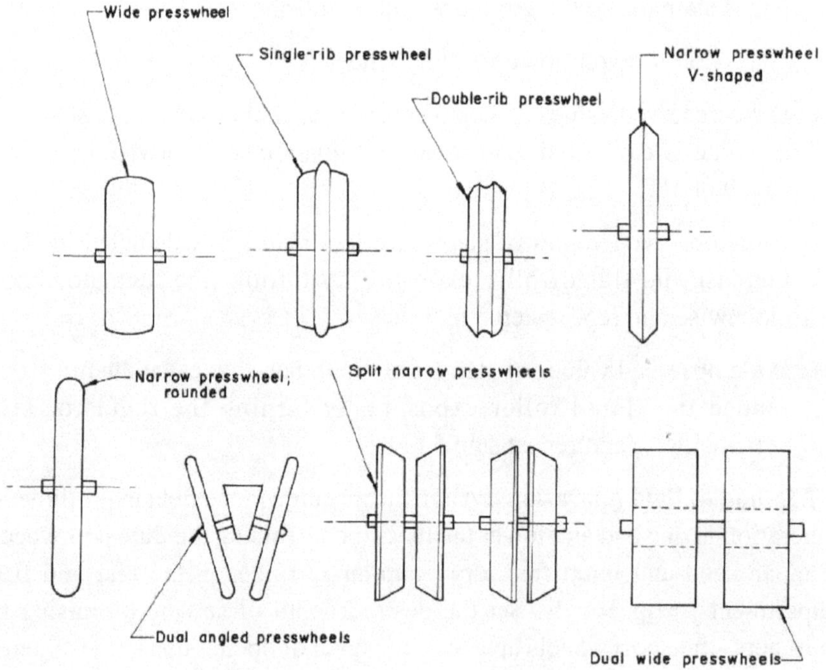

Fig. 4.6: Seed compacting devices used in Seed drills/planters

Lower the ground wheel and operate the drill for short distance to check for smooth rotation of the ground wheel. Check for undue bending of delivery tubes in order to avoid chocking of seed and fertilizer. The horizontal position of the machine should be maintained for uniform penetration of all the furrow openers. At the head lands, raise the ground wheel with help of the level or machine before taking a turn for the next run, to avoid damage to the wheel, shafts and shanks. Do not forget to lower the ground wheel or machine after positioning of the machine for the next run and before starting the run. Also check for any soil sticking to the boots before starting of the run. Furrows are expected to be covered by the natural fall back of the soil; however, if the covering is not proper, under certain conditions, use drag chain or a wooden log tied with ropes at the rear for covering of furrows.

Care and maintenance: Appropriate care of the seed-cum-fertilizer drills/ planters should be taken and be maintained properly to obtain good performance. Observe the following steps for a trouble free service by the machine:

a) Thoroughly wash the fluted roller assembly with water. This is especially necessary for the metering mechanism and box of fertilizer after each day's work. Do not leave any trace of fertilizer either in the hopper or in the metering mechanism in order to avoid corrosion of metallic parts.

b) Lubricate all moving parts such as cut off roller, seed fertilizer rings, bushes, etc except the fluted rollers.

c) Do not carry fertilizer in hoppers, if transporting for long distances, as jerks will pack the fertilizer over orifice of metering units.

d) Keep furrow openers and the hopper painted once in two years.

Trouble shooting guide for fluted roller seed drills are given in Table 4.1.

Seed and fertilizer drills

Drills are used to sow seeds at specified rates and depths continuously in rows. Drills are used for sowing seeds in rows at 15-35 cm apart. This machine sows seeds and fertilizer in separate bands at specified rates in rows at proper depth and covers them with soil. Seed drills may be manually operated, animal-drawn or power operated. Manually operated drills are either single or multi-row seed drills operated by human beings by pushing or pulling it. Animal-drawn seed drills are operated using draught animal power, whereas, power operated drills are seven or more rows with fertilizer attachment, powered by tractor or power tillers.

Table 4.1: Trouble shooting guide for fluted roller seed drills

S. No.	Trouble	Causes	Remedies
1.	Irregular delivery of seed / fertilizer due to skidding of ground wheel	Moving parts need lubrication	Lubricate moving parts
		Hard material stuck in the metering unit	Remove the obstruction
		Too loose soil	Put additional weight on or press the ground wheel.
		Feed plate gap is larger than required*	Reduce the gap*
		Spring is loose/missing **	Replace with new spring **
2.	Seed being damaged	Feed plate gap is less*	Increase the gap*
3.	Inter row variations too large	Metering housing nuts loosened and housing shifted	Secure the housings in position and tighten the nuts.
		Feed plate gaps are not uniform in all the housings*	Maintain uniform gap*
		Rollers exposure is non-uniform**	Fit rollers uniformly**
4.	Once adjusted seed/fertilizer rates vary during field operation	Play in regulator knob mechanisms, feed shafts get shifted due to jerks	Remove the play by inserting thin steel wires in gaps.
5.	Front opener running deeper and rear ones shallow: seeds being not covered	The seed drill is tilting forward	Shift yoke backward on beam, or shift beam on lower holes of hitch, or loosen top link of tractor.
6.	Front opener running shallow and rear ones deeper	The seed drill tilting backward	Shift yoke forward; orShift beam on upper holes of hitch; orTighten top link of tractor
7.	Seed/fertilizer delivery obstructed in tubes	Tubes bent	Straighten tubes
		Obstructions in tubes	Remove straw, stone, mud, etc from opener boots. Do not use wet or lumped fertilizer.

* In case of aluminum fluted rollers.
** In case of nylon-plastic fluted rollers.

Manual seed drills and planters

Manually operated seed drill: The equipment is manually operated hand tool (Fig. 4.7). This is a small manually operated single row seed drill in which fluted roller metering mechanism is provided (Anonymous, 2012; Garg and Singh, 2002). A ground wheel is provided to drive the metering rollers with the help of sprockets and chain. Seed or fertilizer is stored in a small hopper and a long beam is provided by which the implement could be pulled by one operator. Another worker guides the machine. Due to the provision of fluted rollers, it is suited for drilling soybean, maize, pigeon pea, sorghum, green gram, Bengal gram, wheat, rapeseed and mustard etc. Shoe type furrow openers are provided for easy operation. At different flutter roller exposure lengths, desired seed rate in different crops can be achieved. Machine is widely used for inter-row sowing of rapeseed and mustard in sugarcane crop. The machine can sow about 0.04 ha/h with the help of two persons.

Fig. 4.7: Manually operated seed-fertilizer drill.

In another type of manual seed drills wheel hand hoes are used to power the seed metering device (Fig. 4.8). Basically this is wheel hand hoe used for interculture/weeding operation (Garg and Singh, 2002; Pandey *et al.*, 1997). The machine consists of a seed box attached to the mainframe of a hand wheel hoe. A fluted roller assembly is provided at the bottom of the seed box. Fluted roller is rotated with the help of chain and sprockets from the ground wheel. The seed rate can be adjusted with the help of a lever provided on the seed box. The fluted roller used for sowing rapeseed and mustard has 8 flutes. Each flute is 3 mm wide and 2 mm deep. The diameter of the fluted roller is 50 mm and its length, 32mm. For operation, the machine is pulled by rope attached to the hook of machine by one man and other person steers the machine by holding it by the handle. One row and two rows manual drills are available commercially. Working capacity is about 0.04-0.06 ha/h.

a) One row b) Two row

Fig. 4.8: Manual seed drill with wheel hand hoe

Manual garlic/multi crop planter: This is a wheel hand hoe, used for inter-culture operation on which planting mechanism is mounted (Garg and Singh, 2002, Bhardwaj *et al.*, 2004). The planting mechanism consists of a vertical plate with spoons, and receives drive motion from the ground wheel through chain and sprockets (Fig. 4.9). For operation of the planter, a person pulls it from rope attached to the hook and other person steers the machine by holding it by the handle. Upon pulling the planter forward, ground wheel starts rotating transmitting motion to the vertical plate fitted with spoons in the hopper. The hopper is filled with seeds or garlic cloves. As the vertical plate rotates, the spoons pick up the bulb/seed, which is discharged, in the small hopper connected to the furrow opener through a tube. The seed is then dropped in the furrow created by the furrow opener. It is provided with markers for maintaining the row-to-row spacing. Varying the number of spoons on vertical plate can vary plant spacing. This planter is used for sowing of garlic, maize, moong, peas, groundnut etc. Garlic planter is labour saving equipment and requires 90 man-hours to plant one hectare.

Fig. 4.9: Manual garlic/multi crop planter

Animal-drawn seed and fertilizer drills

The simplest seed drill, whicl small farmers use widely, consists o a seed funnel and a vertical seed tub fitted to indigenous plough (Fig. 4.10) The vertical tube is either fitted to th shoe of plough or it is tied with th body to drop seeds just behind th plough in the furrow (Pandey *et al.* 1997). Either the ploughman o another man walks along with the plough and drops the seeds in the funnel. In manually metered se drills, the uniformity of se distribution depends upon the skil man dropping the seeds in funr There is no seed-metering device control the desired amount of se to be sown. There are animal-dra seed drills in use with one; two, thı four and six seed tubes (Figs. 4.11 a 4.12). In another type of country s drills, the fertilizer and seed boxes attached to the soil stirring plough (I 4.13).

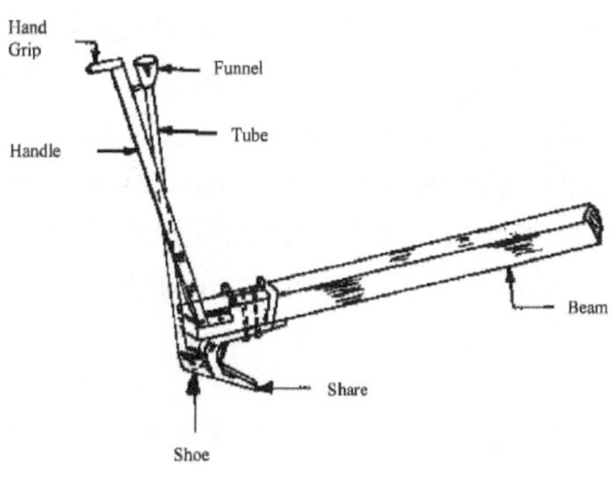

Fig. 4.10: Single row animal-drawn seed drill.

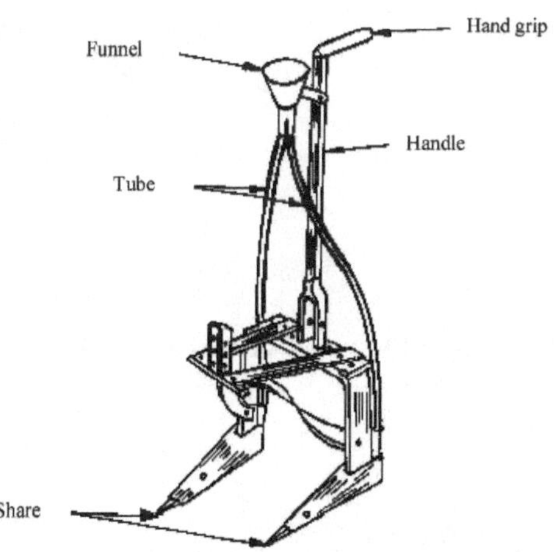

Fig. 4.11: Two row animal-drawn seed drill.

Three-row animal-drawn seed-

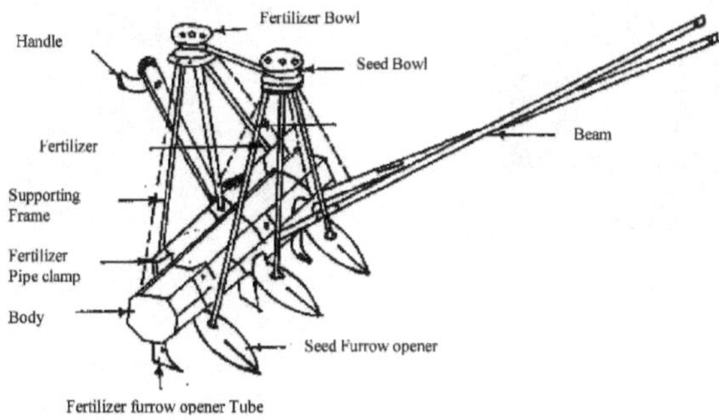

Fig. 4.12: Three row animal-drawn seed-cum-fertilizer drill for dry land.

cum-fertilizer drill has become very popular among the farmers (Fig. 4.14) because it has a fluted roller type metering devices for metering seeds. It is an animal drawn simple, lightweight and compact machine to sow crops like wheat, gram, sorghum, soybean, lentil, pea, sunflower, safflower etc, and drill fertilizer in black soil under rainfed condition. The fluted roller metering mechanism, fitted in the unit, gets the drive from ground drive wheel of 300 mm diameter through chain and sprocket. The shoe type furrow opener with non-clogging boot places the seed at desired depth.

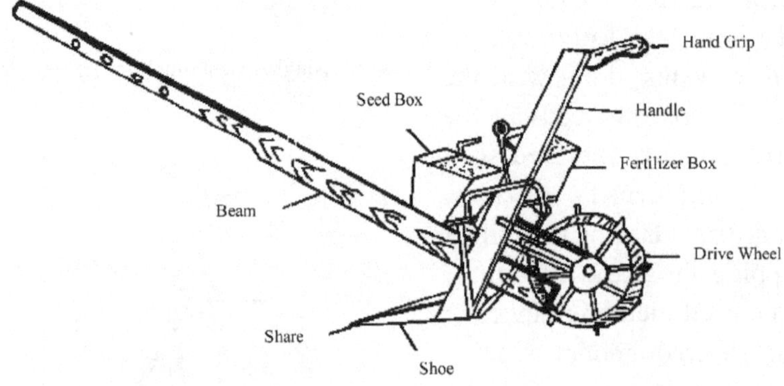

Fig. 4.13: Animal-drawn country plough with seeding and fertilizer attachment

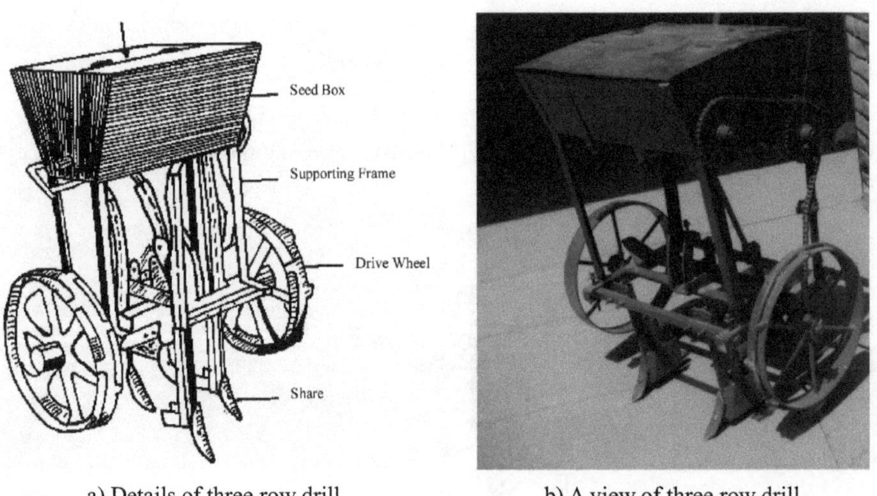

a) Details of three row drill

b) A view of three row drill

Fig. 4.14: Animal-drawn 3-row seed-cum-fertilizer drill

Animal drawn Jyoti multicrop planter: It is an animal drawn implement suitable for planting groundnut, sunflower, safflower, soybean, pigeon pea, Bengal gram, sorghum, wheat and maize (Pandey *et al*., 1997). The equipment can also be used for intercropping. It is a 3-row machine with cell type plastic rollers for seed metering (Fig. 4.15). It consists of separate hoppers for seed and fertilizer. Seeds are metered by a cell type rollers and fertilizer by agitator and orifice in the fertilizer box. Drive is obtained directly from the ground wheel and gears for seed metering and through chain and sprocket for the fertilizer metering. The row to row distance can be easily adjusted and the unit has been accepted well for bold seeds. It saves 75% labour, 30% operating time and 72% on cost of operation compared to conventional method of sowing behind the country plough. It can cover 0.1 ha/h at 70% field efficiency and 2.4 km/h forward speed.

Fig. 4.15: Animal drawn Jyoti multicrop planter

Bullock drawn 4 row groundnut planters: Groundnut is one of the important oilseed crops in India. The farmer has to perform sowing of groundnut at right time and as early as possible to take advantage of favorable situation in dry land conditions. The groundnut farmers are adopting the traditional practice of farm operations with the available bullock drawn implements and labour intensive operations. It is estimated that about 30 to 40 percent production cost can be reduced by introducing mechanization in dry land conditions. The 4 row automatic bullock drawn

Fig. 4.16: Bullock drawn 4-rows groundnut planter during field operation.

groundnut seed drill developed by ANGRAU Hyderabad (Anonymous, 2008 and 2010) consists of seed hoper (8 kg capacity), trough feed type seed metering mechanism, ground wheel of 30 cm diameter and sprocket and chains for power transmission (Fig. 4.16). The width of the seed drill is 120 cm. The row to row and plant to plant spacing are 30 cm and 10 cm respectively. The field capacity varies from 2.0 to 2.4 ha/day.

Power tiller operated seed and fertilizer drills

This machine has been specially designed for operation with a power tiller of 8-10 hp to drill seed and fertilizer in 6 rows with working width of 1.10 m (Pandey *et al.*, 1997). It is provided with depth control gauge wheels and a ground wheel for operating the metering mechanism. Some of its major components are the main frame, seed and fertilizer boxes, metering mechanism, transport wheel, furrow openers, hitch system etc (Fig. 4.17). It is suitable for sowing seeds of wheat, soybean, Bengal gram, sorghum etc in medium and heavy soil. It can cover 0.2 to 0.25 ha/h at forward speed of 2-3 km/h.

Fig. 4.17: Power tiller operated seed and fertilizer drills

There is another machine developed and operated by a power tiller of 10-12 hp and carries out simultaneous tilling and planting operation in a single pass (Fig. 4.18). It can plant two rows simultaneously. It is provided with a pair of depth control gauge wheels, which also serve to activate the metering mechanism. Some of its major components are the main frame, seed and fertilizer boxes, metering mechanism, transport wheel, furrow openers, hitch system etc. It is suitable for sowing seeds of wheat, soybean, Bengal gram, sorghum etc in

Fig. 4.18: Power tiller operated tilling and planting machine.

medium and heavy soil. It can cover up to 90 cm width. The field capacity is 0.08 ha/h at forward speed of 1.5 to 2.0 km/h.

In another power tiller operated tilling and seed-cum-fertilizer drill, preparing the seedbed and drilling the seeds and fertilizer are done simultaneously (Fig. 4.19). The attachment consists of seed box with cup feed metering mechanism, lifting arrangement, seed cut off clutch, tool bar, hoe type furrow openers and seating attachment, all mounted on an easily detachable articulated double wheeled frame. The cup feed seed mechanism is driven by chain and sprockets from the ground wheel. The lifting of tool bar disengages the clutch transmitting power to the seed-metering shaft. The spacing between furrow openers and the depth of sowing are easily adjustable depending upon row spacing desired. This machine is used for sowing seeds like groundnut, maize, sorghum and pulses. It is suited for 10-12 hp power tillers. Operator can ride comfortably while sowing along with the unit.

Fig. 4.19: Power tiller operated tilling and seed-cum-fertilizer drill.

Tractor operated seed and fertilizer drills

The tractor operated seed-cum-fertilizer drill consists of seed box, fertilizer box, seed metering mechanism, fertilizer metering mechanism, seed tubes, furrow openers, seed rate adjusting lever and transport cum power transmitting wheel (Singh, 2007; Singh and Verma, 2009; Pandey *et al.*, 1997). Fluted roller or Bhadson type metering devices are used for metering seeds and gravity feed or corrugated roller type metering devices are used for metering fertilizers. These metering devices are driven by a shaft. Fluted rollers, which are mounted at the bottom of the seed box, receive the seeds into longitudinal grooves of fluted roller and expel them in the seed tube attached to the furrow openers. By shifting the rollers sideways, the length of the grooves exposed to the seed, can be increased or decreased and hence the amount of seed sown is changed. The seed-cum-fertilizer drill is popular in northern region of the country. Seed-cum-fertilizer drills are used for sowing of wheat and other cereal crops in already prepared field. Power is transmitted through chain and sprocket. There could be 7, 9, 11, 13 or 15 rows seed-cum-fertilizer drills depending upon the tractor power available. A tractor mounted 9-row seed-cum-fertilizer drill with fluted feed seed metering mechanism for seeds is shown in Fig. 4.20 and with grooved periphery disc metering mechanism for seeds and fertilizers is shown in Fig. 4.21. Drive wheel is connected through chain and sprocket to seed and fertilizer metering devices. The rate is adjusted by means of a lever provided with seed and fertilizer metering devices. The seed and fertilizer falling through the box are conducted through the transparent tubes and finally dropped into the furrow made by the plough. The field capacity of these drills ranges between 0.2-0.6 ha/h.

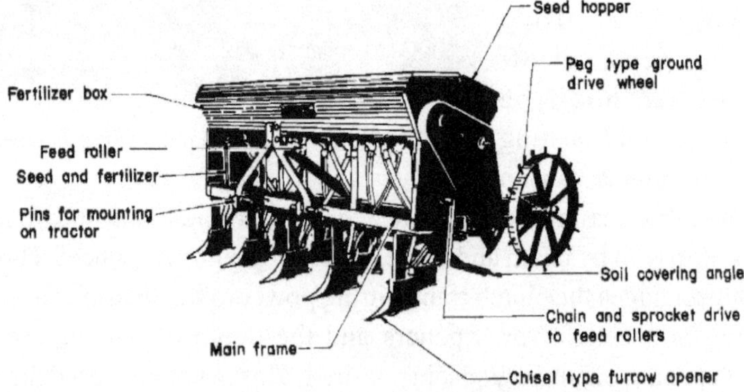

Fig. 4.20: Tractor mounted 9-row seed-cum-fertilizer drill with fluted feed seed metering mechanism for seeds.

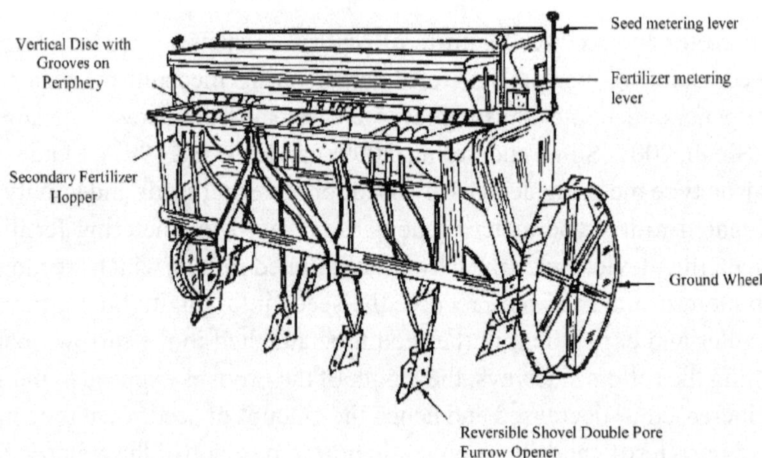

Fig. 4.21: Tractor mounted 9-row seed-cum-fertilizer drill with grooved periphery disc metering mechanism.

Inclined plate planter

In most parts of the country bold seeds are sown by dropping seed manually in a furrow formed by a country plough. After sowing, the furrows are covered using wooden plank. For sowing in small areas dibbling is practiced. In traditional sowing, it is not possible to achieve uniformity in distribution of seeds. There is also poor control over depth of seed placement. The high seed rate and labour requirement are also common problems associated in traditional sowing of bold seeds. For bold seeds such as maize, cotton, groundnuts and pigeon pea, planting and fertilizer placement are the two operations performed by animal drawn planter (Singh, 2007; Garg and Singh, 2002; Anonymous, 2012; Pandey and Ganesan, 2005; Pandey *et al.*, 1997; Singh and Pandey, 2008). They ensure uniform spacing between rows and seed to seed. The animal drawn planter is an important machine in the area for improving productivity. It is animal drawn 3-row sowing equipment suitable for maize, cotton, groundnut, pigeon pea, sorghum, sunflower and pea etc (Fig. 4.22). It consists of main frame, three seed hopper, fertilizer hopper, ground drive wheel, chain and sprocket type transmission and furrow openers, inclined plate with cells type metering mechanism provided for each of the seed hoppers. The seed hoppers are fixed on the tool bars with bushes. The hoppers can slide on these bars for positioning them according to row spacing and allowing the seed delivery point just above the furrow openers (shoe type). The height of the seed delivery spout has been kept close to the ground to achieve seed spacing uniformity. The fertilizer box is provided in front side on main frame with fluted roller metering mechanism. The power to the inclined plate is transmitted from the spiked ground wheel through chain and sprockets and a set of bevel gears. The planter is provided with screw jack mechanism for adjustment of depth and use during transport. The planter is also suitable for sowing of intercrops as different seeds can be filled in different boxes. Drive ratio between ground drive wheel and seed plate can be changed, by selecting appropriate size of sprocket on wheel axle or on counter and main drive shafts. The field capacity of machine is 0.12-0.15 ha/h.

Fig. 4.22: Animal drawn 3-row planter

For sowing bold seeds like groundnut, cotton, maize a planting attachment has been developed which can be mounted behind the tractor operated seed-cum-fertilizer drill (Garg and Singh, 2002). It consists of mainframe, seed boxes, fertilizer box, furrow openers, ground drive wheels, seed tubes, seed metering mechanism and power transmission system (Fig. 4.23). On the rear tool bar of the frame, shoe type furrow openers with modular units of seed boxes are clamped. Each seed box is provided with inclined plate (120 mm diameter) type seed metering mechanism. The seed metering system is driven by a spiked ground drive wheel, fitted on front side of the frame through sets of chain and sprockets and bevel gears. The row spacing (225-450 mm) between furrow openers can be changed by sliding the furrow openers on rear tool bar of mainframe. The field capacity is 0.42 ha/h with an effective width of coverage of 1850 mm.

Fig. 4.23: Tractor mounted 9-row seed-cum-fertilizer drill with fluted feed seed metering mechanism for seeds and planting attachment.

The CIAE 6-row tractor mounted inclined plate planter has been adopted for sowing intercrop (Anonymous, 2008, 2010; Anonymous, 2012). It consists of mainframe, seed boxes, fertilizer box, furrow openers, ground drive wheels, seed tubes, seed metering mechanism and power transmission system (Fig. 4.24). On the rear tool bar of the frame, shoe type furrow openers with modular units of seed boxes are clamped. Each seed box is provided with

Fig. 4.24: Tractor mounted inclined plate planter in operation.

inclined plate (120 mm diameter) type seed metering mechanism. The seed metering system is driven by a spiked ground drive wheel, fitted on front side of the frame through sets of chain and sprockets and bevel gears. The drive ratio (1:1) can be changed at different stages. A common fertilizer box with 6 units for metering granular fertilizer is fixed on the main frame. The drive to fertilizer metering shaft is through the main drive shaft of the planter. The provision has been made for different crops by selecting seed plates and by changing the transmission ratio. The size of cells and inclined plate thickness depends on the seeds to be sown. The row spacing (225-450 mm) between furrow openers can be changed by sliding the furrow openers on rear tool bar of mainframe. The depth control of planter is performed by tractor hydraulic system. The bold grains such as maize, kabuli gram, cotton, groundnut etc can be sown with this machine. The field capacity varies between 0.4-0.6 ha/h depending upon the crop to be planted and row to row spacing.

Raised bed planter

Bed planting system is referred to the planting and cultivation of crops on raised beds. Generally wheat and some other crops are planted on raised beds (Singh and Pandey, 2008; Anonymous, 2010; Anonymous, 2012; Singh, 2007). It has been reported that planting wheat on raised beds improves yield, increases fertilizer use efficiency, reduces herbicides dependence, facilitates better weed management and mobility in the crop field for other cultural operations, less lodging of crops and saves seed, fertilizer and irrigation water. The total production cost compare to flat sown although found reduced marginally at the first planting on fresh beds but it has reduced by 25-35 percent when the beds were reused. Bed planting technique is gaining acceptance by farmers because of more benefit-cost ratio compared to the flat sown crop and it is being assessed for its suitability in different parts of the country for different cropping systems.

In view of specific advantages of bed planting system, raised bed planters have been designed and developed for planting of wheat and other row crops on raised beds (Fig. 4.25). The making of beds in well tilled soil, planting of seeds, basal application of fertilizer and covering and dressing of planted beds are done in one operation. For planting of seeds on permanent beds the same machine is also used for single operation and it adds the advantage of conservation tillage to the bed planting thus reducing the cost of planting compared to the flat sowing of wheat. The raised bed planter is operated in friable soil after 2-3 tillage operations. The shape of the bed is trapezoidal and of furrow is triangular. The size of bed and furrow may vary depending on the soil type and condition of the soil.

The machine consists of 3 mould boards, 2 beds and a seeding mechanism to sow 2 or 3 rows of wheat on each bed simultaneously. The machine can make two beds in single run and the width of each bed is adjustable (35 to 45 cm). A Planting attachment has also been made with the machine for sowing maize, groundnut, cotton etc on the beds. The machine is operated by 34 kW and above tractors. The machine is also provided with a shaper after seeding tynes to shape and compact the beds. The machine can be operated at a forward speed of 2.5 km/h to 3.5 km/h. The field capacity of machine is 0.25 ha/h. The time, cost and operational energy use per hectare of bed planting with preparatory tillage have been found 28.5, 26.2 and 25.0 percent higher respectively compared to the flat sowings.

Two to three rows of wheat are planted at depth of 40 mm on each bed. First irrigation (30 mm) is applied after 3 days of planting for proper germination and emergence of crops. Other cultural operations are similar to that of flat sown crop. Preparatory tillage may be done at friable condition of soil to avoid clod formation and a deeper tillage (150-200 mm depth) is desirable. Sowing on beds by use of bed planter may be done shallow and covering of seeds may be ensured for proper seed-soil contact. Basal dose of fertilizer (P and K full and one third of recommended N) may be applied during sowing and the balance top dressed in two splits. About 75% of the fertilizers recommended for flat sown wheat are sufficient in bed planting for average yield. Application of irrigation water may be frequent depending on soil type and weather conditions but the total water requirement for wheat may not exceed 60-70 percent of what is required for flat sown crop. First weeding may be done by furrowers in furrows after 20-25 days of sowing and subsequent weeding may be done by moving through the furrow spaces.

Bed planting compared to flat sown cultivation of wheat improves yield by 5-10%, saves seed and fertilizer by 25-30%, saves irrigation by 30-35%, prevents lodging of crop, facilitates easy mechanical weeding and reduces herbicides dependence and gives higher benefit-cost ratio. It is more energy efficient and cost effective when planting is done on reused/permanent beds. Two, three and four rows bed planters are commercially available.

a) Two views of raised bed planter developed at R&D institutions

b) Two row raised bed planter *Courtesy*: M/s National Agro Industries, Ludhiana

c) Four row raised bed planter *Courtesy*: M/s Dasmesh Mechanical Works, Amargarh

Fig. 4.25: Tractor operated raised bed planters

Power tiller mounted air assisted seed drill

Fig. 4.26: Power tiller mounted air assisted seed drill in operation. No-tillage seeding machinery

The germination percentage varies widely for small seeds such as sesame and hence the usual practice is to drill these seeds in rows and later get thinned to desired level of crop stand. Though there are number of seed drills developed in the country for handling different types of seeds, most of these equipment are found unsuitable for very small seeds. The metering precision has been found to be inversely proportional to the seed size. Air assisted drill is one machine, which is capable of drilling any size of seed to the desired seed rate (Pandey *et al.*, 1997). A typical air assisted drill has an air powered distribution and delivery system consisting of a powered blower and a distributor head. It has an air plenum into which seeds are fed, conveyed and finally distributed to individual furrows by the distributor head. The machine consists of an air blower, distributor head, seed pipe, seed hopper with fluted roller seed metering mechanism, lifting arrangement, seed cut off clutch, tool bar, hoe type furrow openers and seating arrangement (Fig. 4.26). All these are mounted on an easily detachable double-wheeled frame. The tail wheels are carried on separate telescopic cross frame. The fluted seed metering mechanism is driven by a chain and sprocket from the tail wheel. A manually operated lever arrangement lockable by a cable mechanism lifts the tool bar mounted with four hoe type furrow openers. The shanks of the furrow openers are fixed on the tool bar through square clamps. The spacing between the rows can be adjusted from 300 mm for 4 rows to 600 mm for 2 rows. The machine covers 0.25 ha/h. Lifting the tool bar can cut off the power to seed metering shaft. Thinning should be done after 15 days of sowing for all crops to achieve required plant spacing.

No-tillage seedling machinery

The functional requirement and operation environment of a No-tillage seed drill/planter is considerably different from a conventional planter used in prepared field. The basic differences are the no-tillage drill/planter normally used in fields full of straw and sod. The drill should be able to cut and manage the straw without adversely affecting the seeding operation. No-tillage drill works in an unprepared field which requires more force for furrow opener penetration. Covering and firming of the soil around seed is critical and difficult in unprepared field. The drill should have suitable device for these functions. Unit power requirement for no-tillage drill is normally higher than conventional one. Uniformity of seed and fertilizer placement is affected when metering mechanism is powered by ground wheel drive mainly due to undulations in the field. As a consequence, minimum tillage or No-tillage planting of crops is being adopted (Garg and Singh, 2002; Pandey *et al.*, 2006; Shukla *et al.*, 1984, 1990, 1996, Anonymous, 2010, Singh and Pandey, 2008; Pandey *et al.*, 1997).

Minimum tillage is defined as reducing the field trips by machinery for seed-bed preparation to a minimum number which not only helps clear the path for the seeder/planter but also allows pre-emergence herbicide to enter the soil and be more effective. The term no-tillage designated a procedure where the crop is planted directly into a chemically drilled sod or crop residue with no prior mechanical seedbed preparation. Total soil disturbance is limited to that required for proper placement of the seed. No-tillage is a form of conservation tillage keeping soil erosion to a minimum via conservation of the crop residues at the surface. No-tillage saves time in seedbed preparation; the yields are higher or often equal to that of conventional tillage; the improved porosity of the surface soil resulting from aggregation enhances infiltration of water and improves aeration; results in reduction of expenditure on seed-bed preparation; packing of soil is reduced due to reduction in number of trips of machinery over the field; reduces evaporation losses & minimizes short term drought effectively; helps in decreasing surface run-off & superior soil and water conservation through increased infiltration of rainfall & thus results in increasing water use efficiency; improves soil structure & reduces erosion of soil and results in saving of fuel. The main disadvantages are: residues in the field may reduce the soil temperature, which delays in germination, emergence and early growth of the crop sown; better control of weeds is not obtained; not suitable for all crops and it results in more power requirement for planting the crop.

Zero till drilling is referred when seeds are drilled directly into the uncultivated soils just after the harvest of previous crop (mainly rice) by eliminating the tillage operations. It is a nine, eleven or thirteen-row unit consisting of fluted rollers for metering of seeds and vertical rotors over adjustable openings / variable hole-mesh type for metering of fertilizer. The ground drive wheel supply power through sprocket and chain for metering of seed and fertilizer. The furrow openers are of inverted 'T' type spaced at 200 mm row spacing. Major components of zero till drill are frame, furrow openers, seed and fertilizer boxes, seed metering device, fertilizer metering device, power transmission unit, depth-control side wheels, hitch point and wooden/iron platform.

Some researchers have suggested minimum soil disturbance as a design consideration for no-till seeders while other reports insist on some form of strip-tillage to ensure proper seed placement and reduce possible phytotoxicity from crop residues. Seeding tools promoting considerable soil surface disturbance have been blamed for weed problems. When all the concepts incorporated in the many designs of zero-till planters are compared, one of two approaches is always used: minimal disturbance approach (drills developed in this approach attempts to open a clean slot in the soil and to seed without tilling the seed zone), and micro-seedbed approach (widely used). In micro-seedbed approach a narrow tilled zone is created and seedling into this zone with a conventional row crop seeder is done. For creating tilled zone, fluted coulter, rotary attachment or some other device, may be used. A 5-6 cm wide and 7-10 cm deep soil manipulation is fully adequate. Furrow openers are the most important component of direct drilling machines. In many commercial machines furrow openers are individually mounted and spring loaded for better penetration and ground contact. One important aspect of the furrow openers performance is that straw should not be left in contact with the seed and the furrow sides should not be smeared. Following types of furrow openers are used on direct drilling machines (Shukla *et al.*, 1984).

Single disc with coulter attachment: The coulter is used for cutting straw and giving a precut to the soil. The diameter of the discs in commercial machines normally varies from 450-500 mm. In one machine, single disc is slightly offset both vertically & horizontally. This "bursts" the slit & gives more tilth below the surface. It also allows the plow shaped seed coulter to be placed behind the disc, almost touching it. This share completes the brushing of the slit, deposits the seed, and then sweeps the trash out of the slit.

Double disc furrow opener with coulter attachment: A triple disc coulter assembly consists of a vertical pre disc followed by two flat discs angled towards each other at approximately 10 degree included angle and touching the bottom. The leading disc cuts a slot, which is opened by the following pair of discs. Seed and fertilizer are individually dropped between the rear discs into the slot. The front disc is easily adjusted so that it can be kept 15 mm deeper than the rear discs. The rear discs are mounted on a specially designed heavy-duty casting incorporating seed and fertilizer tubes. The triple disc can cut from near surface for grasses to a deep 76 mm for cereals.

Hoe coulter assembly: It consists of a 200 mm diameter vertical disc followed by a narrow "V" shaped hoe coulter. The hoe coulter assembly creates a "V" shaped groove and it also creates loose rubble on the surface. The hoe coulters have limitations with respect to moisture retention in dry soil conditions.

Chisel coulter assembly: The assembly consists of a flat 250 mm diameter vertical pre-disc followed by a hollow chisel coulter with small near-horizontal sub-surface wings. The chisel coulter assembly creates minimal surface disturbance and produces considerable sub-surface shattering. Its groove are roughly the inverse of the hoe and triple disc coulters in that it is narrow at the top & wider at the bottom.

Inverted 'T' type furrow opener: Inverted 'T' type furrow opener is being used in some drills developed in New Zealand and adopted at G.B. Pant University of Agriculture and Technology, Pantnagar (India) and in Pakistan. The furrow opener is mounted on spring tine in original machine. The furrow opener creates soil micro-environment well suited for seed germination, can drill in both tilled and untilled soil and can handle to some extent residue or trash left from previous crop.

Animal drawn, power tiller operated and tractor operated no-till drills are commercially available (Fig. 4.27). Power tiller operated zero-till drill is preferred in hilly region. In Himalaya region the farming is practiced in terraces and sloppy land. The soil resource has suffered from degradation over the past many years. The major contributor to this trend in soil organic matter loss is the tilling of soil with disc plough and cultivator in preparing the land for seeding. If we are to counter the effects of soil degradation caused by excessive tillage of the soil, we must find and adopt new methods of annual cropping. In zero tillage crops are planted with minimum disturbance of soil by placing the seeds in a narrow slit 3-4 cm wide and 4-7 cm deep without land preparation. A prototype of power tiller operated zero-till drill has been designed for sowing wheat crop. The effective field capacity is 0.09-0.10 ha/h at a forward speed of 2.1-2.2 km/h with field efficiency 56-62%.

Tractor operated zero-till seed-cum-fertilizer drill has been developed to sow wheat directly in rice-harvested fields without preparing the seedbed. It is a 9/11/13-row unit consisting of fluted rollers for seed metering and agitators over adjustable openings for fertilizer metering. The ground drive wheel supplies power through sprockets and chain for metering of seed and fertilizer. The furrow openers are of inverted 'T' type, spaced at 200 mm (adjustable). The field capacity of the machine is 0.3-0.4 ha/h with about 75% efficiency. Use of zero-till drill for direct sowing of wheat after rice was found to be advantageous in terms of 50-60% saving in time and 40-50% saving in cost of sowing as compared to the conventional practice of seed bed preparation and sowing with seed-cum-fertilizer drill.

a) Animal drawn no-till drill

b) Power tiller operated no-till drill

c) Tractor operated no-till drill
(*Courtesy*: M/s. National Agro Industries Ludhiana)

Fig. 4.27: Animal drawn, power tiller operated and tractor-operated No-Till Drill

Strip-till-drill

Tillage is one of the major farm operations and is an important contributor to the total cost of production. Excessive tillage is energy and time consuming and costly operation. It is considered harmful to the soil structure. It also contributes to wind and water erosion. The rising cost of hydrocarbon fuels which are bound to be exhausted sooner or later, availability of herbicides coupled with the motive of timely sowing and reducing the cost of production has provided enough incentive to the researchers all over the world to investigate tillage operations more closely. The first approach to the development of a suitable minimum till drill was to develop an appropriate attachment to an existing drill for adoption of the technology. With this end in view, different types of commercially available seed cum fertilizer drills were tried for direct drilling of wheat in manually harvested paddy fields. This was intended to assess the problems arising from the use of commercially available drills under no-tillage regime. The major problems encountered were: accumulation of straw and stubble in front of the tynes, formation of clods, poor coverage of seed and fertilizer leading to bird-damage to the seeds, excessive slippage and lack of contact of ground wheel due to uneven fields leading to skips in the placement of seed and fertilizer, and higher power requirement for the operation. To overcome the above-mentioned problems, an attachment in the form of rotary blade in front of furrow openers of the existing seed drill has been developed (Garg and Singh, 2002; Shukla *et al.*, 1990, 1996).

The strip drill is essentially a 9 row seed-cum-fertilizer drill with a rotary blade attachment for minimum soil manipulation running ahead of the normal furrow openers (Fig. 4.28). A tractor of 35 or higher horsepower (26.11 kW) operates it. The rotary attachment consists of a frame with a rotor having '9' flanges. Each flange has 6 C-type tines (blades) which prepare a 75 mm wide strip in the front of every furrow opener. Thus with every row, 125 mm of the strip is left untilled and only 40 percent of area is tilled. Tilling and sowing is done simultaneously. The spacing between the flanges is the same as the row spacing for the crop to be planted. Power to the rotor shaft is provided from the tractor PTO through a speed reduction gear box and chain and sprocket drive. The rotor revolves at a speed of 300-rpm corresponding to the rated PTO speed of 540 rpm. The rotary attachment is provided with an MS sheet cover to protect the power transmission system. It also helps to reduce the soil cover over the seed. Strip till drills are used for sowing wheat after paddy without any prior seedbed preparation. It can save 50-60% fuel and 65-75% time as compared to conventional method. The effective field capacity of strip till drill is 0.3-0.4 ha/h. There is vast potential for increase in yield due to timeliness of sowing. Under the minimum tillage system, the diesel fuel consumption in planting operation only (no separate seedbed preparation required) is 18 l/ha while diesel fuel used for seedbed preparation and sowing under conventional tillage system is 60 l/ha. Thus, there is a saving of diesel to the extent of 42 l/ha.

Fig. 4.28: Tractor operated strip-till drill

Roto till drill

The roto seed drill is a combination of a rotavator and a seed drill which gives a new dimension to the sowing operations being utilized in the country (Garg and Singh, 2002; Pandey *et al.*, 1997). The roto seed drill doesn't need field preparation before sowing and therefore next crop can be immediately sown after the harvesting of previous crop. The two most important components of the roto seed drill are rotavator and seed drill (without tines). It is extensively used for sowing a wide variety of crops like maize, wheat, pea, mustard etc. It has a huge demand in the market for durability, performance and less energy consumption. It is a low maintenance seed drill which is easy to operate and handle. The various components of the roto seed drill are frame, seed and fertilizer box, seed tubes, seed metering mechanism, fertilizer metering mechanism and rotavator (Fig. 4.29). The frame is made of high quality steel with suitable braces and brackets so that the frame is strong enough to withstand all types of loads. Because of varying needs of cost, complexity, weight distribution, stiffness, power output and speed, there is no single ideal frame design. The seed and fertilizer box is made of mild steel or galvanized iron with a suitable cover. The purpose of the seed tube is to deliver the metered seed rate to the surface.

Fig. 4.29: Roto-till seed drill
Courtesy: M/s. National Agro Industries Ludhiana

The roto seed drill works with the fluted feed type metering mechanism. In this the adjustable fluted rolls are present to collect and deliver the seed into the seed tube. Traction wheel for operating seed and fertilizer metering mechanism is provided on side. The fertilizer metering is generally done by two methods namely, cell feed mechanism and auger feed mechanism. The cell feed mechanism is a mechanism in which seeds are collected and delivered by a series of equally spaced cells on the periphery of a circular plate or wheel. The auger feed mechanism is a distributing system consisting of an auger which cause the fertilizer to flow evenly in the field, through an aperture at the base or on the side of the hopper. Many of the fertilizer drills of the country had an auger feed mechanism. The rotavator is a P.T.O driven implement. The main purpose of roto seed drill is to reduce the number of land preparation operations to be performed, as it uproots the stubbles of crop residue like sugarcane, cotton, paddy etc and chopps them into small pieces and mixed them back into the soil which helps to build up the humus content of the soil. The rotavator performs all the above mentioned functions in one operation thereby saves fuel and time of the farmer. Rotavator is equipped with a gearbox and driven forward, or held back, by its wheels. The gearbox enables the forward speed to be adjusted while the rotational speed of the blades remains constant which enabled the operator to easily regulate the extent to which soil is engaged. The roto seed drill is customized according to the requirement of the customer but it is manufactured in three standard sizes namely 1.52m (5 feet), 1.83 m (6 feet), and 2.13m (7 feet). The machine can cover 0.25 to 0.4 ha/h.

Happy seed drill

Until recently, direct drilling of any crop into combine harvested rice stubbles from a reasonable rice yield has not been possible without prior burning or removal of straw. To solve the problem, a new machine called as 'Happy Seeder' capable of direct drilling wheat into heavy rice residue loads, without burning in a single operation by managing only that part of straw which is coming just in front of furrow openers has been developed by Punjab Agricultural University, Ludhiana in collaboration with CSIRO, Land and Water, Griffith, Australia. Happy Seeder combines the stubble mulching and seed drilling functions into the one machine (Anonymous, 2013). Happy seeder (HS) consists of a rotor for managing the paddy residues and a zero till drill for sowing of wheat (Fig. 4.30). Flail type (Gamma) straight blades are mounted on the straw management rotor which cuts (hits/shear) the standing stubbles/ loose straw coming in front of the sowing tine and clean each tine twice in one rotation of rotor for proper placement of seed in soil. The rotor blades/flails guide/push the residues as surface mulch between the seeded rows. This PTO driven machine can be operated with 45 hp tractor and can cover 0.3-0.4 ha/h. It has been observed that wheat yield in case of happy seeder is nearly 10% more than conventional sowing practice. Weed matter is nearly 50% lesser on happy seeder plots compared to conventionally sown plots. Happy seeder sows wheat directly in paddy residue in combine harvested field hence, prevents residue burning thus reduces air pollution. Mulched crop residue improves the soil health and adds organic matter to the soil. Soil temperature regime is conducive for adoption of long duration wheat varieties due to increased window period. Substantial water saving has also been recorded due to avoidance of first irrigation and mulching.

Courtesy: Kamboj Mech. Works Amritsar, and M/s Dasmesh Mechanical Works Amargarh

Fig. 4.30: Happy seeder

Tractor operated cotton planter (Inclined plate with cell)

Cotton as a crop as well as a commodity plays an important role in the agrarian and industrial activities of the nation and has a unique place in the economy of our country. Cotton popularly known as "White Gold" is grown mainly for fiber. In addition to this, cotton seed is second important source of the edible oil. India has been a traditional home of cotton and cotton textiles. The domestication of the cotton cultivation for clothing of humanity is considered to begin in Asian sub-continent using diploid cottons. India is the only country where all the four cultivated species of cotton are grown. Our economy is consistently influenced by cotton through its production and processing sectors, and by generating direct and indirect employment to more than eight million people. India is the largest cotton growing country in the world, where 60 million people are impacted by cotton. Several factors that influence the germination of seed and the stand obtained are the quantity of seed planted, viability of seed, treatment of the seed with chemicals to kill soil micro-organisms, use of fuzzy seed, use of delinted seed, planting depth, type of soil, moisture content of the soil, types of seed dropping mechanism, size of cell in planting plate, keeping seed hopper filling depth uniform, distribution of the seed, type of furrow opener (runner or shovel), width of furrow opener, prevention of loose soil getting under seed, uniform coverage, type of covering device, pressing or firming the soil around the seed, type of press wheel or device, placement of fertilizer in relation to seed at planting time, time of planting in relation to season, water standing in furrow after planting and the experience and skill of the operator.

Cotton planter places a desired quantity of seed, without scattering, at regular intervals within the rows. Therefore, metering devices for the cotton planters holds a desired quantity of seed per hill and drop them on the ground at equally spaced intervals. Several types of metering devices are used in cotton planters, namely, vacuum disk type, inclined plate type, belt type, vertical rotor type, and roller type. Of these, the inclined plate type seed metering mechanism is most widely used in cotton planters. The scattering of seed results in unevenly sown hills, which require an additional work to thin or transplant later. It is generally known that the scattering is caused mainly by the improper design of the metering device, seed tube and seed release height. Straight and short seed tubes reduce the scattering.

Cotton planting is conventionally done by seed drill and other planters. Planting of cotton and other bold seeded crops are generally done by planters where row to row as well as plant to plant distance is essential for the seed saving and better crop growth. The seeds are sown in line at the depth of 30-40 mm with two seeds per hill maintaining the desired spacing between row and plants. In order to get desired plant population/ crop stand by other than this method, extra seed rate is used which requires thinning of crop. This not only requires extra seed rate but also increase cost of labour for thinning and extra time for desired crop establishment in field. Precision planting is proper placement of seed in row at desired depth and at equal intervals. Precision planting saves seed and fertilizer to best advantage and increase yield by enabling good cultivation practices. This technique results in uniform plant spacing, seed depth and helps further mechanization of intercultural farming operation that reduces the total cost of cultivation. In inclined cell plate type planter seed sowing produces more consistent row to row distribution of seeds and reduces plant stand variability (Anonymous, 2010, 12; Anonymous, 2008a and 2010a; Singh, 2007; Pandey *et al.*, 1970).

The machine consist of a main frame, ground wheels, seed and fertilizer hoppers, furrow openers, power transmission system, and three point hitch (Fig. 4.31). The metering of Bt. cotton seed is done with the help of inclined plates with cells on periphery (Fig. 4.31). The seed metering mechanism is lowered down near to ground level to avoid scattering of seeds. Ground wheel provides drive to the metering shafts through the sprocket and chain arrangement. The depth of planting can be adjusted by lowering or raising ground wheel. The seed rate used is 1.5 to 2.5 kg/ha. The average field capacity of Bt. cotton planter is 0.7 to 0.8 ha/h. Bt. cotton planter saves costly inputs like seed, time, labour and money, better quality of work in less time as compared to traditional practice, placement of seed and fertilizer at proper depth and spacing and the slit opened by the furrow opener is narrow thus moisture remained conserved for longer period to have better germination even at high temperature.

Fig. 4.31: A view of Bt. Cotton planter in operation.

Check row planter for cotton

Traditionally the planting is done manually in check rows. The check row spacing ranges from 90 to 120 cm depending upon the period of planting, crop variety and type of soil. The present method is laborious and time consuming. In order to meet the need of precision in planting, attempts have been made to develop such a machine (Pandey *et al.*, 1997). The parameters taken into account while designing the machine are: i) at a time 2-3 seeds should be planted on the check row hill, ii) about 10 g of granular fertilizer is applied to each hill and iii) it should be simple and easy to repair. The check row cotton seed planter consists of main frame, furrow opener, seed box with metering device, fertilizer box with metering device, power transmission system and marker (Fig. 4.32). The main frame for mounting seed boxes and furrow opener is made from mild steel box section. Three spear headed furrow openers are mounted at spacing of 90 - 120 cm. It is supported at the front soil-working end by 10 mm rod fixed on a horizontal square pipe. Three seed boxes are mounted on these furrow openers with seed metering plate fitted in aluminium housing. Each seed plate has one cell, and driven by straight spur gears mounted on a common shaft. Three trapezoidal fertilizer boxes are fitted beside the seed box on the same shaft. The single grooved roller made of nylon material is used for metering fertilizer. Power from the ground wheel is transmitted directly to the common shaft for metering seeds and fertilizer. An eccentric weight is provided on the ground wheel to maintain cell orientation at head lands. To maintain spacing between successive rows marker are fitted at the end of main shaft. It gives an effective field capacity of 0.613 ha/h.

Fig. 4.32: Tractor operated check row planter for cotton

Tractor operated multi crop planter for seed spices

Among the major 20 seed spices cumin, fenugreek, fennel, coriander and ajwain are the seed spices largely cultivated in India. The cultivated land share of seed spices for Rajasthan is 43% in country. The sowing of seed spices is mainly done by broadcasting method or drilled in small plots at a spacing of 25-30 cm and depth of 1-1.5 cm. Looking to the need of farmers and export potential of seed spices a 5 row planter with individual hopper boxes has been developed and is further modified to 7 row multi crop planter with seed metering mechanism mounted on common frame (Anonymous, 2008 and 2010). The power transmission is through chain and sprocket and drive wheel (Fig. 4.33). The seed metering mechanism of star wheel (plastic) is made of circular rotor of 90 mm diameter with 10 cells of 20 mm length. The height of seed dropping from hopper is kept at 40 cm to get the accurate placement of seeds at depth between 10-15 mm. The inverted T type furrow opener of smaller size was fitted in comparison with other commercial drills. The machine has also fertilizer drilling attachment and variable row to row spacing arrangement. The machine is suitable for sowing of cumin, coriander and fenugreek. The seed rate of 6.5 - 7.5 kg/ha, 12-15 kg/ha and 9-10 kg /ha is observed for cumin, fenugreek and coriander respectively. The field capacity of machine is 0.28-0.3 ha/h with depth of seed placement as 12-15 mm. Looking to the requirement of small farmers a two row hand operated multi crop seed spices planter has also been developed for sowing of fenugreek crop.

Fig. 4.33: Seed spices planter

Maize (corn) planter: Maize planters available are either bullock operated or tractor operated. Bullock operated planters are usually single, two or three row planters (Fig. 4.34), whereas tractor-operated planers are 2 to 4 row or even more (Fig. 4.35). Trailing and tractor-mounted maize planters can be classified according to the manner in which seeds are dropped: drill or hill-drop. Four types of seed plates are used for planting maize viz. edge-drop, flat-drop, flat-drop round-hole and full-hill plate. Planters are usually adjustable for row spacing as well as plant spacing. Precise planting rate is possible in maize because of size and nature of seed.

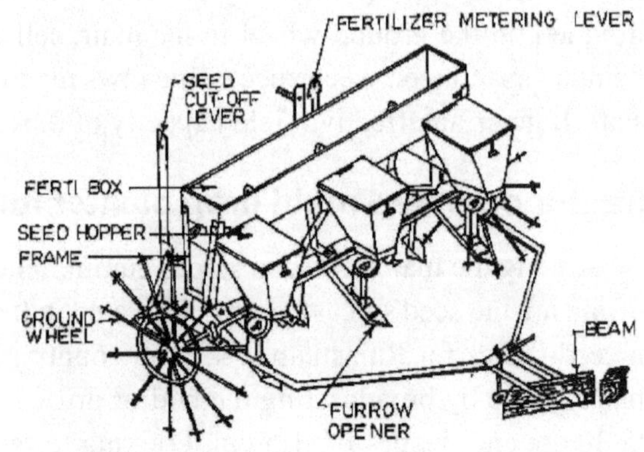

Fig. 4.34: Animal-drawn 3-row planter with fertilizer attachment.

Accurate single seed selection of graded seeds with properly fitted seed plates is possible. Maize is sown on both flat land and ridges. It can be check-rowed also for cross cultivation. This practice facilitates weed control, especially in very wet years. Mostly it is a horizontal or inclined plate type of planters and single seed selection can also be made. Planting rates depend on the number of cells per plate and the drive rates. At higher speeds, scattering of seed is pre-dominant due to bouncing. The accuracy of a planter depends upon the uniformity of kernels, shape of hopper bottom, speed of plate, shape and size of cell and fullness of hopper and height of seed hopper from ground. A cone-shaped hopper bottom causes the seed to gravitate into the cells. The height of the planter box from the ground level should not be more than 45-60 cm.

Maize is a traditional kharif season crop. However, it can be grown successfully in Rabi season also by adopting suitable variety as well as agronomic practices. Planting of maize in Rabi season is recommended on the southern side of east-west ridge to harvest maximum sunshine (Singh *et al.*, 1985; Singh *et al.*, 1988). Planting of maize manually on ridges made either manually or by tractor operated ridger requires about 100-125 man-h/ha. A planting attachment to a 3-bottom tractor mounted ridger can reduce labour requirement substantially (Fig. 4.35). A ridge planter for winter maize consists of a 3-bottom tractor mounted ridger, planting mechanism, furrow openers and spring loaded ground wheel. The distance between ridger bottoms can be adjusted as per requirement. Planting mechanism is of inclined plate type and has two units. Planting mechanism gets drive from spring loaded ground wheel, which moves in furrow. The machine has two furrow openers of reversible shovel type. These furrow openers are mounted on a frame, which could be moved horizontally by an arrangement consisting of a set of levers and handle. This handle can be operated by tractor operated while sitting on the tractor at the end of each turn so as to place furrow opener on the southern side of east-west ridges. The furrow openers make furrows of 3-4 cm depth on the southern side of ridge and seeds are dropped in the furrows through transparent seed tubes. The machine maintains seed to seed spacing of 20 cm but can be adjusted as per need. It can cover about 1.5 ha per day or more depending upon machine width and number of rows planted.

a) Two row ridge planter for winter maize

b) Four row maize planter

Courtesy: (Kamboj Mech. Works, Amritsar)

Fig. 4.35: A tractor operated 2-row and 4-row maize planter

Tractor-operated pneumatic planter

Tractor-operated pneumatic planter consists of main frame, aspirator blower disc with cell type metering plate, individual hopper for each row, furrow openers, PTO driven shaft and ground drive wheel (Fig. 4.36). The machine works on the air suction principle and a suction pressure of 2 kg/cm^2 is developed by the machine. Air is sucked through a rotating plate having various holes radially. For cotton, seed plates with hole angle of 90°, hole size 3 mm and with 4 holes are used. A seed coming in contact gets stuck to the holes on the plate and falls when suction is cut-off at the lowest position near the ground. The fall of the seed is synchronized with the predetermined seed spacing. Since the seed is lifted under suction no mechanical damage takes place. Row to row spacing is adjustable. The drive to the metering mechanism is given through the ground wheel by means of chains and sprockets. The machine is suitable for sowing cotton, soybean, groundnut, sorghum, pigeon pea, maize etc by changing the seed plates. The field capacity of the pneumatic planter is 0.49 ha/h and field efficiency 77%.

A 8-row vegetable pneumatic planter (Fig. 4.37) was imported for direct seeding of small seeds (Anonymous, 2010). With the machine it is possible to plant seeds at 200 mm row spacing with average seed to seed distance of 94 mm and at an average 27 mm depth, in 8 rows along the beds. The average field capacity of the machine was 0.2-0.25 ha/h.

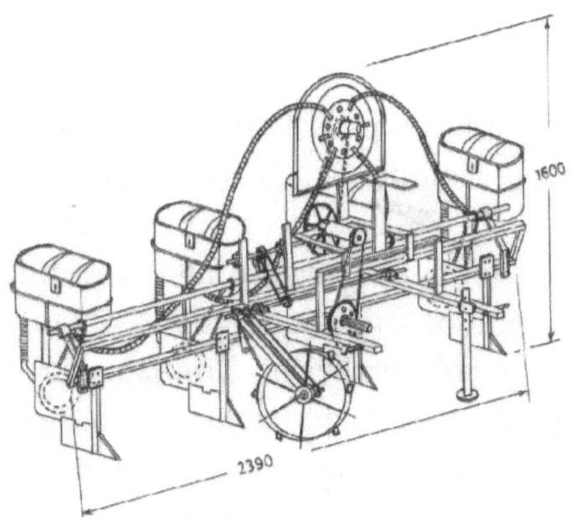

a) Isometric view

b) First prototype

c) Modified prototype

Fig. 4.36: Tractor operated pneumatic planter

Fig. 4.37: Tractor operated pneumativ/vegetable seeder

Sugarcane planting equipment

Sugarcane is one of the most important cash crops of India. It is grown in most part of the country in total area of about 5 million ha. This crop needs the highest labour as compared to any other crop. With the economic growth of the nation, governmental welfare measures for unemployed youth, providing employment guarantee at village level, arduous field operations, dignity involvement etc; timely availability of required number of farm labour has become a deterrent factor in carrying out required field operations in time. Due to increased labour wages, cost of production of sugarcane has gone up, resulting in lowering down the profit margin of cane grower. Many times grower feels distressed in carrying out field operations for want of manpower. Thus, mechanization is the only option left to grow sugarcane economically and gracefully. Lot of efforts has been made at Indian Institute of Sugarcane Research (IISR), Lucknow and elsewhere to develop suitable equipment for sugarcane farming. Some of the efforts have proven their utility at farmer's fields, in terms of cost effectiveness, reduction in labour requirement, timely operations, providing dignity to labours etc. Sugarcane provides raw material for sugar, khandsari and jaggery industries. It's by-products and other items are used for production of alcohol, paper and many other products. Of the total production of sugarcane about 48% is utilized for sugar, 40% for jaggery and khandsari and about 12% for seed, feed and chewing purposes.

Sugarcane needs deep and well prepared soil. Planting of sugarcane comprises many operations such as opening of furrows, cutting of cane into pieces known as seed setts, placement of setts, fertilizer and insecticide in the furrows and providing soil cover over the setts. Furrows are opened with the help of tractor drawn ridgers or bullock-drawn ridgers. Forty to forty five man-days are required in one hectare to carry out other operations. Arranging such a huge number of labour in a day is very difficult. This results into moisture loss of soil as well as seed setts. Institutions in India has developed various models of sugarcane planters viz. semi-automatic, automatic and later cutter planters suited for tractors as well as bullocks (Singh, 2014; Yadav, 2003; Singh and Sharma, 2008; Shukla *et al.*, 1978). The different variants of tractor operated sugarcane cutter planters are either tractor PTO or ground wheel driven. Sett cutting is continuous and uninterrupted in PTO driven planters but proper sett metering can only be achieved at a particular forward speed and PTO rpm combination. Sett metering remains same in ground wheel driven planters but precaution is required that ground wheels do not skid and remain in firm contact with soil. Multipurpose sugarcane planter has also been developed which not only works as three row cutter planter but can also be used for intercultural operation and earthing-up in cane field. There is saving of more than 60 per cent in the cost of planting operation by using sugarcane cutter planter as compared to traditional method.

Sugarcane has physiological phenomena of top dominance i.e. the top buds which are comparatively young, succulent and immature germinate first and try to suck food from the whole cane, if it is not cut into pieces. This adversely effects the germination of lower buds. This phenomenon is called "Top dominance". To overcome this problem the cane is cut into setts. Three budded setts are recommended for planting but at many places 2 budded and single budded setts are also used. Sugarcane is propagated through its cuttings called setts. Its planting consists of two operationally different but sequentially connected operations of sett preparation and later, their placement in furrows. Both these operations are labour intensive and get delayed when there are shortages of labour at peak planting times. For both these operations machines have been developed. They differ in designs according to requirements. Method of sugarcane planting differs from place to place in two ways viz. one on the method of planting setts in the soil and other on the spatial arrangement of one sett from the other. In the first case 5 methods are employed i.e.

Flat planting - In this case shallow furrows of 10-12 cm are made, setts are planted and covered fully with the soil and the field becomes flat. This type of planting is common in lighter soils.

Furrow planting - In this method of planting ridges and furrows are made of 15-20 cm depth. After planting the setts in the furrows, they are covered with about 3-5 cm soil leaving the furrows partially open, which helps in economizing in irrigation water. This type of planting is common all over the country. In this case the setts are planted end to end, 25% overlap, 50% overlap and in some cases 100% overlap. Normally 30,000 setts (3 budded) are used in one ha. The seed rate varies from 5-7 tonnes/ha.

Trench planting - This is similar to furrow planting except that the trenches are made 30 to 45 cm deep. It is not common in India.

Pit planting - In this case circular or square shaped pits of 45-60 cm diameter or length and 15-25 cm depth are dug at 90 or 100 cm apart from both side and in each pit 3-4 or even more setts are planted. Ten to twelve thousand pits are required to be dug per ha. This method is also not common in India.

Wet planting in furrows - This is similar to furrow planting above. Furrows and ridges are made using tractor/bullock operated ridgers. Irrigation water is allowed in the furrows and setts are dropped in the wet soil and pressed in the mud.

Sugarcane planters

Sugarcane planting requires 6 operations to be done simultaneously namely opening of furrows, dropping of setts in the furrows, applying fertilizer, applying fungicidal/insecticidal solution on the setts and soil surface, covering of setts and slightly pressing of the soil cover. Machines have been developed to do all these six operations simultaneously in one pass. The sugarcane planters can be classified into two broad categories i.e Drop planter or Cutter planter. In the first category cut setts are planted in the furrows while in the second category the setts are also cut in the same machine. Drop planters are more common as it provides an opportunity of selection of individual setts and requires comparatively less power. Cutter planters are heavy but reduce one separate operation of sett preparation. They are common in Australia. In both types of machines semi-automatic and automatic versions are available. Different types of sugarcane planters animal drawn and tractor operated have been developed in India.

Semi-automatic bullock drawn sugarcane planter

This machine is a single row unit for flat planting of sugarcane in ligher soils (Fig. 4.38). It consists of a three wheeled fore-carriage drawn by a pair of bullocks and a trailed implement attached behind it. The fore carriage carries a seat for operator and 2 seed boxes on either side of the seat. The trailed unit consists of a rectangular metal box open at top and bottom and made up of a share point in front for making furrow. This box is rigidly attached to a short beam in front and is provided with a handle at the back. The free end of the beam is hinged to a vertical

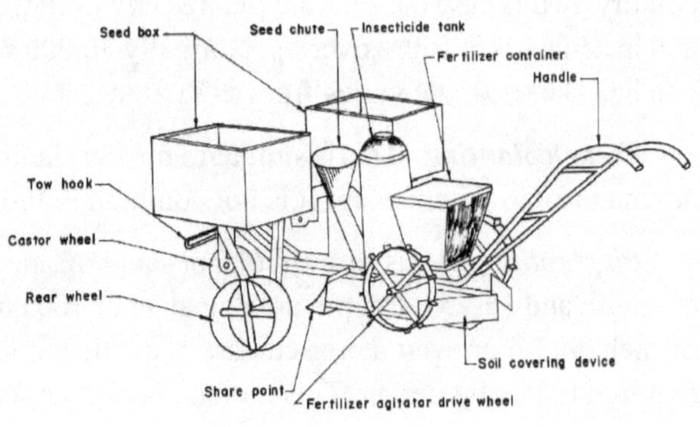

Fig. 4.38: Animal drawn sugarcane planter

clevis in the front of fore carriage. A fertilizer box and chemical tank are mounted on the trailed unit. A covering device is also fixed at the back of the trailed unit. Three persons are required to operate the machine - one for guiding the bullocks, another for dropping setts and third for guiding the implement. It covers about 1.5 ha in 8 hours at 90 cm row to row spacing.

Semi-automatic tractor operated sugarcane planter

This machine is similar in operation as above (Fig. 4.39). It is used for furrow planting for which ridger bodies are used. The machine is made in units each requiring about 15 hp. Depending upon the hp of the tractor 2 or 3 units can be used at a time. Each row unit gives a coverage of 1.25 to 1.5 ha/day. Two row units are more common. The dropping of setts is done manually. Other 5 operations are done by the machine. The output of this machine is between 2.5-3.0 ha/day. The cost of planting is about 50% that of conventional method.

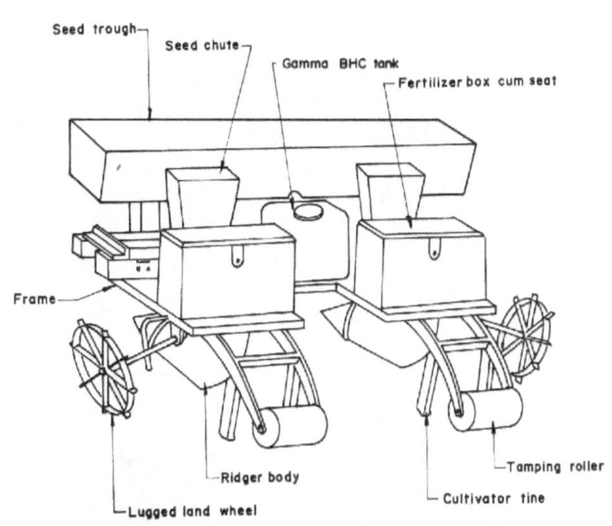

Fig. 4.39: Semi-automatic tractor operated sugarcane planter

Roto drum type tractor drawn sugarcane planter

It consists of a frame, two furrow openers, two seed chutes, two vertical rotating drums each having 12 compartments, seed container, fertilizer hopper, power transmission system, wheels and two seats (Fig. 4.40). The diameter of the ground wheel is 75 cm. Vertical rotating drums in which setts are filled, are also known as roto drums and has a diameter of 29 cm. Height of bottom of the feed drum from the ground is 73.0 cm. The lower end of chute is 9 cm above the ground. It has a ridger type furrow opener. The covering device is two tines for each furrow and a roller. Machine is operated by a 35 hp tractor and has two rows. In this machine, two persons sitting on the machine lift the setts from hopper and fill the compartment of roto-drums by sugarcane setts. The hopper is filled with setts before starting the operation. The moving roto-drum drops the setts in the feeding chute, which place them in the furrows. Also chemical and fertiliser are applied side by side before covering the setts. Machine can cover about 0.2 ha/h at a tractor speed of about 2 km/h. Performance of machine is satisfactory and it can save 40-50% labour and 15-20% cost of operation. Eye damage is 1.5-3%.

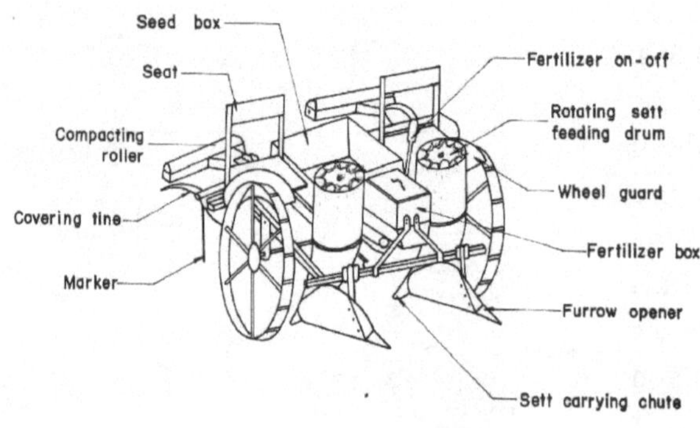

a) Details of machine

b) A view of machine

Fig. 4.40: Roto drum type tractor drawn sugarcane planter (Shukla *et al.*, 1978)

Tractor operated sett cutter planter for sugarcane

It is a whole stick sugarcane cutter planter and consists of ridger body attached to frame and create furrows, sett cutting unit, fertilizer application unit, chemical application unit, sett covering unit and seed box (Fig. 4.41). The planter is tractor mounted with three-point linkage system and is hydraulic load free. It carries more quantity of pesticides, insecticides, whole stick sugarcane to ensure covering more field area without refilling at every turn. Cutting of the cane into 350 mm length is done automatically in the machine and sett are treated with insecticides at the cutting point. The setts are placed in the furrow created by ridger bodies automatically with overlapping up to 30%. The pesticide is also sprayed on the ends of sett in the furrows. The seed placed in the furrow is covered immediately after the treatment, the furrow is closed and rows are leveled by the leveler provided in the machine. The machine is PTO driven. The field capacity is 0.20 ha/h.

Fig. 4.41: Tractor operated sett cutter planter for sugarcane (Singh, 2014; Yadav, 2003)

Tractor operated single row and two row adjustable sugarcane cutter (ground wheel driven) planter

This unit has provision to plant sugarcane in a single row. The planter can operate with the help of 25 hp tractor (Fig. 4.42). The cutting mechanism of the machine is operated by ground wheel. This planter unit has been designed and developed by Vasantdada Sugar Institute, Pune and ICAR under NATP project on Sugarcane Mechanization. The VSI-NATP has also developed a two row tractor operated type machine with an adjustment of 150, 120 and 90 cm row-to-row spacing (Fig. 4.43). The Patta system of sugarcane planting (mostly adopted in Maharashtra) is also done by this planter. The planter is operated by 45 or higher hp tractor. In order to achieve uniform length and better cutting of the setts the rotary cutting along with the cylindrical feeding (positive feeding) mechanism is provided. The placement of sugarcane setts is adjustable as per the requirement i.e. end to end, overlapping and maintaining suitable spacing between the setts.

Fig. 4.42: VSI-NATP single row sugarcane cutter (ground wheel driven) planter (Singh, 2014; Yadav, 2003)

Fig. 4.43: Two row adjustable tractor operated sugarcane cutter planter (Singh, 2014)

Tractor operated multipurpose implement for sugarcane

Sugarcane planting is very labour intensive job and involves considerable human energy. Planting creates the foundation for a crop and plays an important role in its growth and yield. Being costly machine it has been desired to develop multipurpose implement, which could facilitate interculture and earthing up operations in addition to planting of sugarcane setts. The equipment with all the three attachments has been developed. The machine consists of a frame, feeding chute, coulter, ridger, seed box for sugarcane, seat for the operator, covering and packing device. It also has a provision for attaching, cultivator, seed drill and puddler (Fig. 4.44). In three-row machine the operator lifts the cane from seed tray and puts the same in the slanting chute. Cane slides down and a cutter cuts the cane into setts automatically. When cane comes to end another cane is placed in the chute. In this way, there is some free time to operator. The three persons feeding cane in to the cane cutting units feel comfortable and appreciated the new mechanism. This machine can cut whole sugarcane into pieces of uniform size, place in the furrow prepared by ridger, setts are covered by soil and then soil is pressed in one operation. Fertilizers and chemical can also be applied. The field capacity of machine is 0.20 ha/h. For land preparation the equipment can be used as tractor operated nine-tine cultivator. The equipment in interculture mode has six standard tynes with reversible shovels that provides soil cover in three rows of cane placed in opened furrows. Three extra tynes with shovels facilitates interculture operation in three inter rows of cane in single pass. The field capacity in interculture mode is 0.72 ha/h. For earthing up, three-rows of cane can be covered by this equipment with a field capacity of 0.66 ha/h. For puddling, beater type sub-unit is mounted with the main frame of the equipment having the field capacity of 0.35 ha/h.

a) Sugarcane Planting b) Interculture

c) Seed drill mode d) Puddling

Fig. 4.44: Tractor operated multipurpose sugarcane cutter planter attachment (Singh, 2014; Yadav, 2003)

Paired row planting is also gaining popularity because of convenience in intercultural and tying operations. Farmers can also grow companion crops in wide space. Planting of cane in two rows are carried out simultaneously at spacing of 30 cm (Fig. 4.45). Another paired row may be planted at 120 cm or 150 cm apart depending upon the need (Singh, 2014). The Indian Institute of Sugarcane Research Lucknow has developed an improved heat treatment technology for disease free cane seed. It is known as Moist Hot Air Treatment system and the equipment used to carry out the treatment is called Moist Hot Air Treatment Plant (MHAT), Singh (2014).

Fig. 4.45: Tractor operated paired row sugarcane planter

Tractor operated twin-auger-digger for sugarcane planting

Ring pit method is another method of cane cultivation. Though, cane yield is very high in this method of cane cultivation but production cost is also very high, mainly because of digging of pits and placement of setts etc. A tractor operated pit digger (2 bottoms) has been developed (Anonymous, 2008). The machine can dig two pits at a time. The machine consists of a heavy duty frame made from a channel on which two rotary units having doubled helical auger blade (EN-32) has been fixed in opposite direction around a high pressure circular pipe of 98 mm diameter (Fig. 4.46). A set of three triangular shaped blades made of EN-45 is fitted at the end of each helical auger for digging of soil. The helical auger is made from a sheet of 5 mm thickness. The distance between the two augers is 1220 mm. The power from the tractor PTO is transmitted to the auger through a reduction gear of ratio 22:12 and from gearbox to augers through V-belt (C-87) and pulley consisting of three grooves. The effective diameter and depth of the pit is 68-70 cm and 28-35 cm respectively. The capacity of the machine varies from 0.020-.025 ha/h and the fuel consumption was from 5.5-6.0 l/h. Despite of mechanical digging, the practice takes a lot of time and labour for putting the FYM and sugarcane setts in star and soil for covering the setts in the individual pits. The increase in yield is about 12.4 percent. The labour requirement and cost of operation respectively in pit planting is about 65% and 230% more as compared to general practice of ridge sowing. Also some the other problem likes inter cropping and the agronomic experts anticipated uniformity of distribution of nutrients. The machine can dig a set of two pits in 40-50 seconds depending upon the depth of operation including time loss from one set of pit to the next (Fig. 4.47). Under average conditions it is capable of digging 140-180 pits per hour. Due to small spacing between the pits, some of excavated soil falls back into the pit when the auger is lifted up.

Fig. 4.46: Tractor operated twin-auger-digger

Fig. 4.47: tractor operated twin auger digger during field operation
Courtesy: Dasmesh Mechanical Works Amargarh (Punjab)

Sugarcane planter-cum-seeder

Sugarcane is wide spaced crop. Generally, in sub-tropical India, cane rows are spaced at 75 cm. Two types of machines have been developed which can be used for planting of sugarcane as well as companion crops like wheat, urd, moong, masoor etc (Singh, 2014). The IISR sugarcane planter-cum-seeder has been developed for flat planting of sugarcane and drilling two to three rows of companion crop seeds in between two rows of sugarcane (Fig. 4.48). This machine can cover one hectare in four to five hours.

Fig. 4.48: IISR sugarcane planter-cum-seeder

RBS cane planter

This is a tractor driven machine and performs many operations in single pass i.e. complete planting of cane in two furrows at 75-90 cm apart, with 4.5 cm soil covering over setts placed about 20 cm deep in furrows, makes two raised beds with about 23 cm top width and drills seeds of companion crops like wheat, urd, moong etc along with fertilizer in two rows on raised beds. The machine is known as RBS Cane Planter (Fig. 4.49). RBS stands for raised bed seeder. The machine is very useful. Farmers can plant sugarcane and companion crop in this system. Five to six persons cane accomplish the complete job in about five hours in one hectare area. Under this method furrows are made at 80 cm spacing and three rows of wheat are drilled on raised beds. Depth of furrows remains approximately 22 cm and top width of raised bed 52 cm. Three rows of wheat seeds are drilled on this bed at 17 cm spacing. Wheat crops grow on raised beds and sugarcane setts are dropped in the furrows. This system is cost effective including substantial savings in seed, fertilizer and irrigation water.

Fig. 4.49: IISR raised bed seeder-cum-sugarcane planter (Singh, 2014)

Sett cutting machine

In case of conventional planting, sugarcane setts are prepared manually by local tools. Sett cutting machines are also available for preparing single, double or three bud setts. Traditionally, setts are prepared by sugarcane cutting knives. About 30,000 three budded setts/ha are required for planting which require about 15-20 man days to prepare setts. A sett cutting machine has been developed at IISR, Lucknow and PAU Ludhiana (Srivastava and Sharma, 1977; Pandey *et al.*, 1997) for cutting setts which can be operated either by a 5-7 hp engine, electric motors or even by a tractor on low throttle (Fig. 4.50). With a team of 4 persons - 2 for cutting the canes and 2 for help in lifting the canes; an output of approximately 12-13 thousand setts/h are obtained. The cost of sett cutting is about 1/3 to 1/2 that of traditional system. The machine consists of two circular saws mounted on a platform. Suitable guards are provided over the blades for safety. The canes are fed in bundles (3-5 at a time). Cut setts are dropped into a fungicidal tank from where they are removed and planted.

Fig. 4.50: Sugarcane sett cutting machine

Tractor operated garlic planter

Planting season of garlic coincides with harvesting of Kharif and sowing of Rabi crops. Manual planting of garlic involves 400-500 kg/ha of garlic cloves. Development of a tractor operated garlic planter with provision for other crops of the region such as maize, groundnut and soybean has been developed. To enhance the capacity a tractor operated garlic planter has been developed (Anonymous, 2013a). It has star wheel type seed metering mechanism for both seeds and fertilizer (Fig. 4.51). It can cover 12 to 17 rows at a time. The observed seed rate varies from 500 to 700 kg/ha mainly dependent on size of garlic cloves. The spacing of garlic cloves ranged from 5 to 10 cm. The field capacity of garlic planter is 0.51 ha/h.

Fig. 4.51: Tractor operated garlic planter

Tractor operated cumin planter

The sowing of cumin seed is mainly done through broad casting method in fields. The seeds are either broadcasted or sown in line by iron or wooden hooks at 25-30 cm apart and seed is dropped manually in lines. Later on seed is covered by a sickle or broom as sowing depth is kept as 1 to 1.5 cm (Anonymous, 2013a). A five row planer has been designed with individual hopper box and seed metering mechanism mounted on common frame with transmission through chain & sprocket connected through ground drive wheel (Fig. 4.52). The seed metering mechanism is of star wheel (plastic) type made of circular plastic rotor of 98 mm diameter rotor with 10 cells of 27 mm length and 6 mm width. The height of hopper has been kept as 45 cm to get the accurate placement of seeds in shallow furrows. Inverted 'T' type of

Fig. 4.52: Tractor operated cumin planter

furrow openers is used to get the shallow depth of seed placement. The machine can be used for sowing of cumin seed besides fenugreek and coriander. Seed rate of 10-12 kg/ha for Cumin,18-20 kg/ha for Fenugreek and 10 kg/ha fo Coriander is obtained. The field capacity is observed as 0.44 ha/h and depth of placement between 1 cm to 2 cm.

Potato planter

Potato is an important cash crop, which can be cultivated in wide variety of soils and weather conditions. One of the main bottlenecks in increasing the area under potato cultivation has been its high labour requirement for planting, earthing up and inter-culture. Potato crop yield also responds to improved soil aeration and deep tillage. Potato planting is done in rows of 50-60 cm with 15-20 cm plant-to-plant spacing. Potato planter performs the function of furrow opening, seed metering, seed placement at proper depth and forming of the ridges to cover the seed tubers. It is observed that manual planting and earthing of potato require 300-320 man-h/ha. Non-availability of adequate labour during the planting season necessitates the mechanization of potato planting. Potato planters that are available and in use are animal drawn, power tiller operated and tractor operated units of automatic and semiautomatic type. Animal drawn and power tiller operated planters are single row where as tractor mounted are 2 or more rows. The latest designs are equipped with fertilizer as well as herbicide applicators (Singh, 2007; Singh and Verma, 2009; Pandey *et al.*, 1997; Verma *et al.,* 1992). Potato planter mainly consist of a seed hopper, a metering device for controlling the seed spacing along the row, furrow openers, a covering device to the ridge, drive mechanism, frame and hitching system. To drive metering mechanism a separate ground drive wheel is provided. Hoppers are normally rectangular with sloping bottom or sides. An auxiliary hopper is provided at the bottom of the main hopper where the picker wheel or metering mechanism is mounted. For manual dropping of seeds, a trough is provided near operator to feed potato tubers to conveyor belt or rotating magazine. Potato tubers move downward under gravity and movement depends upon the size shape and weight of tubers, angle of rolling and angle of repose. Adjustable baffles are provided in the hopper to adjust the size of outlet. It can saves 40-50 % labour requirement and 15-20 % cost of operation in comparison to traditional method.

There are different types of metering devices used to regulate the potato seeds such as belt cup type, magazine type and picker wheel type. Magazine type metering device is rotated in the horizontal plane in an open housing

(Fig. 4.53). Operator sitting in front of rotating magazine keeps on filling the seed from trough. Rotating magazine carries the seeds to seed tube, which in turn places it in the furrow. Seed spacing within the row is maintained by maintaining proper speed of magazine. Belt cup type metering mechanism is used in semi-automatic potato planters (Fig. 4.54). The cups are fastened to an endless belt. Belt moves horizontally and passes through hopper, picks up seed in cups as it moves and carries it to the seed tube. During this process some of the cups remain unfilled, which are filled by the operator. The equipment is operated by a 35 kW tractor. It consisted of a frame, furrow openers, seed box, two revolving magazines or two belt-conveyer with cups, seats, seed tubes and a ground wheel for transmitting the power to the shaft. The diameter of wheel can be varied by adjusting length of lugs on the wheel. The machine is mounted on a 3-point linkage of a tractor. Two persons sitting on the machine fill the seeding cups by picking tubers from the hopper. In rotary magazine type planter, cups are filled manually with any

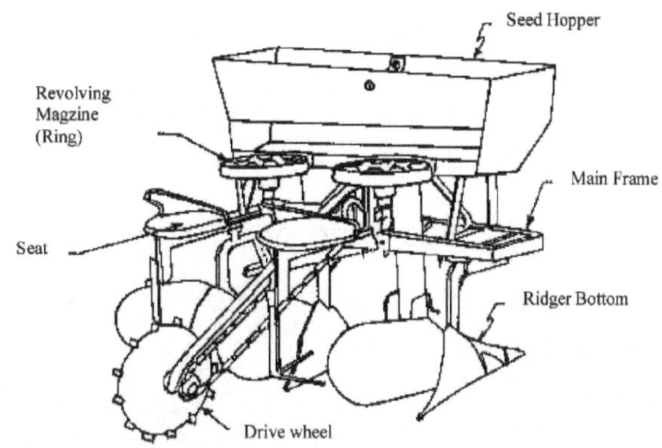

Fig. 4.53: Tractor operated potato planter with revolving magazine type metering device.

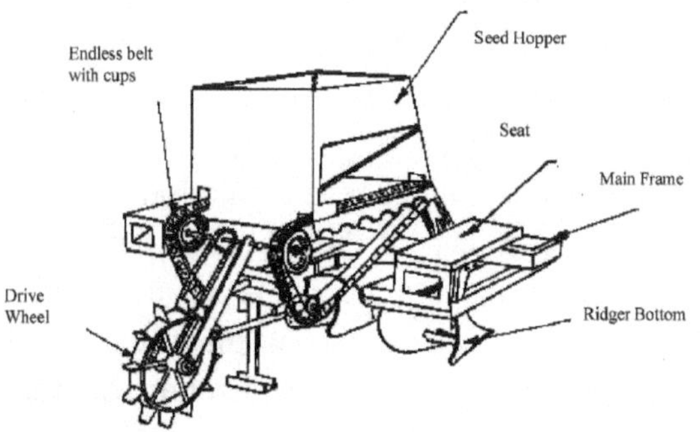

Fig. 4.54: Tractor operated potato planter with belt-cup type metering device.

size of tubers. However, in case of belt-conveyer type machine graded tubers are used and 70-80% of cups are filled automatically and remaining manually. Row to row and plant-to-plant spacing can be adjusted. Working capacity is 0.15 to 0.2 ha/h.

Power tiller operated potato planter

A power tiller operated semi-automatic potato planter is suitable equipment for sowing potato, applying fertilizer and making ridges in single operation (Pandey *et al.*, 1997; Agrawal *et al.*, 2003). A rotating device is used for placing the potato and metering rollers for applying fertilizer (Fig. 4.55). Depth of operation is adjustable up to 80 mm by fixing furrow opener standard up & down through nuts and bolts. The power to metering mechanism is drawn from ground wheel through sprocket and chain arrangement. The machine is simple in construction, light in weight and easy to operate on terraces. The field capacity is 0.03 ha/h.

Fig. 4.55: Power tiller operated potato planter.

Tractor operated automatic potato planter

The automatic potato planter consists of a hopper, two picker wheels for picking the tubers, seed tubes, furrow openers, three bottom ridger to form two ridges, a fertilizer metering system and a frame (Fig. 4.56). The hopper for potatoes is made of MS sheet. It is rectangular in shape at the top with sides sloping towards bottom. At the bottom of the hopper there are agitators to improve the delivery of potato tubers to feeder. The rear wall of the hopper has window and a gate for feed regulation. The agitators and the screw conveyors ensure proper feed of the tubers to the picker unit. The picker units of the planter are set in motion from the ground wheel of diameter of 52 cm (adjustable). The spacing between adjacent plants can be varied by suitably modifying the gear ratio. The planter has fertilizer distributors for the application of fertilizer during planting. It is used for planting potato on the ridges without involving laborers. It saves about 70% labour and 40-60% cost of operation. Field capacity is 0.3 ha/h.

Fig. 4.56: Tractor operated automatic potato planter
Courtesy: Droli Industries (Regd) (BASANT), Moga

Vertical belt paired row potato planter

Tractor operated vertical belt paired row potato planter consists of two vertical rubber belts fitted with metal cups in paired rows for picking potato tubers, provision for fertilizer application and shovel type furrow openers (Fig. 4.57). Arrangement of cups on the belt is in Zig-Zag manner. Number of cups per belt is 40. Planter is equipped with multispeed gears to adjust the seed spacing. Provision to adjust the row spacing from 609-711 mm and also to spread the manure on both the sides is also provided. Field capacity of the paired row planter is 0.24 ha/h at an average forward speed of 2.5 km/h.

Fig. 4.57: Tractor operated vertical belt paired row potato planter (Anonymous, 2013)

Paddy transplanter

Paddy is the most important cereal crop and staple food in most of the South East Asian countries. Its production is maximum per unit area of land amongst the cereals and is adaptable to a wide range of agro-climatic conditions.

India is the second largest rice producing country in the world. The area and production of this crop is expected to increase further as the paddy cultivation has become more paying with the introduction of high yielding varieties. Paddy is grown generally by transplanting or direct sowing. In spite of considerable progress made in the area of direct seeding of paddy with chemical weed control, the transplanting remains the most common method of paddy sowing. Transplanting essentially refers to the planting of 20-35 days old and 20-30 cm long seedlings under wet land conditions. Transplanting has a number of advantages over direct sowing like the time that crop occupies the land is reduced by 2-4 weeks resulting in less water requirements, helps the plant a better start over the weeds and hence, lesser growth of weeds, selection of more vigorous seedlings and less seed requirement.

The usual practice of transplanting manually is to hold a bunch of seedlings in one hand and 1 to 2 seedlings separated by the other hand are fixed in the puddled field in bending posture, which affects the spinal cord of the workers. Transplanting of paddy manually requires high labour requirement (about 200-250 man-h/ha) and still the work done is not satisfactory. Also, scarcity of labour during peak season of transplanting delays the operation, which causes progressive decrease in the yields. Therefore, there is a need for mechanical transplanter (Singh, 2007; Garg, 1992; Garg *et al.*, 1982; Garg and Sharma, 1985; Garg and Sharma, 1984; Garg, 1987; Singh and Pandey, 2008, Pandey *et al.*, 1997).

Transplanters are classified on the basis of nursery i.e. machines using washed root seedlings and machines using mat type seedlings: In the machines using washed root seedlings, the seedlings having 4-6 leaves and 20-30 cm high are removed from the traditional nurseries and are washed well with water. In some cases, the roots and tops are pruned to reduce entwining to facilitate separation and picking. These seedlings are then neatly arranged on the seeding boxes of the machine for transplanting. Thus more labour is required for nursery uprooting, washing and arranging in the boxes (about 175 man-h/ha). On the other hand mat type seedlings are raised on the polythene sheet with the help of frames. Soil mixed with farmyard manure and fertilizer is filled in the frames and known quantity of pre-germinated seed is spread uniformly in these frames. Seedling mats are developed after 20-25 days, which are uprooted and used on the machine. The soil thickness in the mats is only 1.5-2.0 cm. Transplanters using mat type seedlings are becoming more popular due to their superior performance and reduced labour requirements (about 50 man-h/ha). Transplanters are classified on the basis of power requirement i.e. manually operated, self-propelled walk behind type, self-propelled riding type and tractor operated.

Mat type seedlings are not currently grown in India because of manual transplanting. Detailed procedure for raising mat type seedlings is described below:

a) Select the nursery location in such a way that the transport of the seedlings to the field where transplanting is to be done is minimum. Nursery field should have a good soil. About 50 sq.m area is sufficient for sowing nursery for one acre.

b) Prepare the field by harrowing and leveling at about 10% moisture content. It is better to add few baskets of farm yard manure in the field before harrowing.

c) Spread 50-60 gauge polythene sheet of 90-100 cm wide over a prepared level ground. It is better if the sheet has perforations.

d) Place 3-4 steel frames having 14 compartments of 40 x 20 x 1.5 cm in each frame over the polythene sheet. About 350 g of polythene sheet spread to a length of about 20 meters in sufficient for preparing seedlings for one acre. Number and size of compartment can vary according to machine specifications.

e) Fill soil in the frames uniformly up to the top surface. The soil is lifted from both the sides of the frames by making shallow furrows/ channels. Soil for filling in the frames can also be prepared separately by mixing one fourth quantity of farm yard manure. Care must be taken at the time of sowing that there are no stones in the soil which is filled in the frames.

f) Spread about 700 g of pregerminated seed evenly in each frame to achieve seedling density of 2 to 3 plant/cm^2 in the mats. About 10-kg seed is sufficient to sow 200 mats required for transplanting one acre.

g) Cover the seeds by a thin layer of soil and sprinkle water by hand sprayer for proper setting of the soil

h) Lift the frames and put these at the next place

i) Repeat the above procedure for sowing required number of seedling mats. Two persons can sow seedlings for 3-4 acres in a day. For proper irrigation, it is better if 7-8 frames are grown on a single bed.

j) After sowing, irrigate the field by flooding so that the seeds are not submerged or washed away. Care must be taken that the seedlings mat area is always wet.

k) Spray or broadcast the seedlings after about 10 days at the rats of 300 g urea/200 mats after irrigating the nursery.

l) The seedling mats become ready after 25-30 days of sowing. Drain the water from the nursery field a few hours before nursery uprooting.

m) The seedling mats are uprooted by giving a cut with a sharp blade/sickle along the boundaries of the mat. Lift the nursery mat and dress it to the proper size of the tray of the transplanter, if needed. One person can uproot the seedlings for 2-3 ha in a day. The uprooted mats are transported to the field either by tractor trailer or by any other means.

Nursery sowing seeder

For uniform distribution of seed a nursery-sowing seeder has been development. The seeder has a diameter of 19 cm and width 100 cm which is equal to the width of the nursery sowing frame and is made of GI sheet (Fig. 4.58). A long shaft passes through the center of the seeder and the ends of the shaft are connected with the wheels. The seeder is also connected with a handle for its movement. The seeder has 2 partitions and in each compartment, two

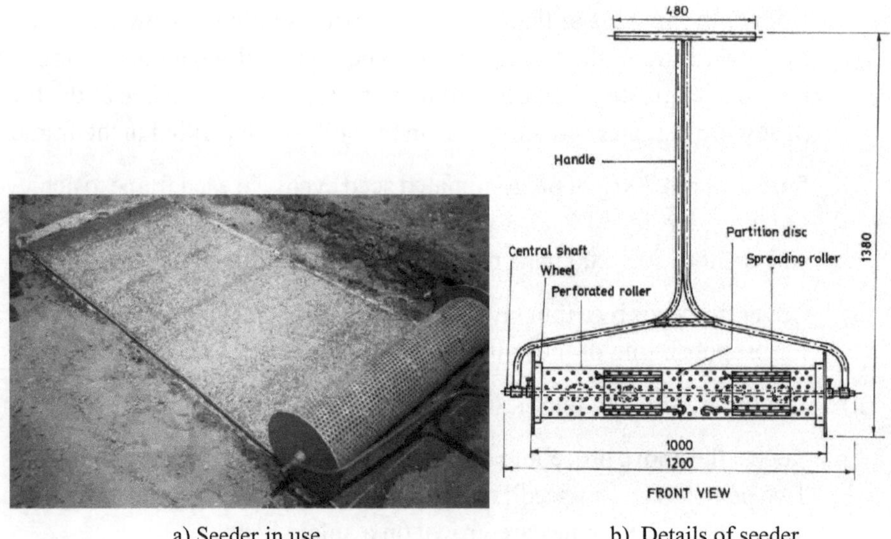

a) Seeder in use b) Details of seeder

Fig. 4.58: Nursery sowing seeder

agitators have been provided on the shaft for agitating the paddy seed. This has been done to reduce the chocking of seed at the opening holes. The seeder has the openings of 1 cm diameter on full length of the roller for uniform distribution of the seed. The holes are staggered for better distribution. Two feeding gates have been also provided on the seeder for filling the seed in each box.

Manually operated 6-row rice transplanter

It is a manually operated rice transplanter suitable for transplanting of mat type rice seedlings in puddled soils in rows. The manual rice transplanter consists of floats made of marine plywood, seedling tray that accepts the mat type nursery, a tray indexing mechanism and pickers for planting seedlings (Fig. 4.59). When the operator pulls the machine and operates the handle, the picker, six in number, gathers two or three seedlings and place them in the puddled soil. The row-to-row distance can be maintained at 200 mm. It can cover 0.04-0.05 ha/h and labour requirement is 40-50 man-h/ha.

Self-propelled rice transplanter

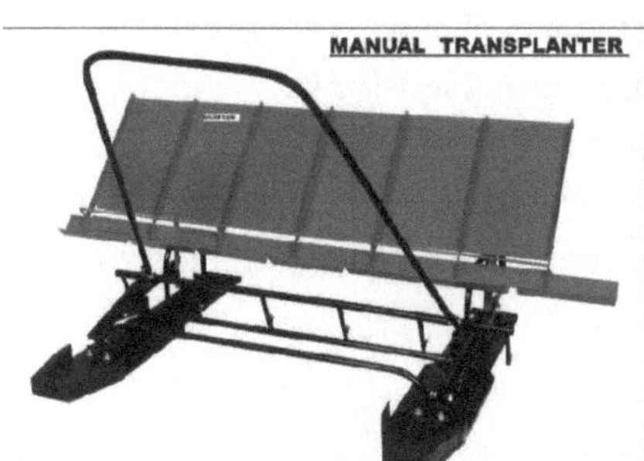

Fig. 4.59: Manually operated 6 row rice transplanter

The self-propelled rice transplanter is provided with single wheel (lugged steel wheel for field and pneumatic wheel for transport) and driven by 2.94 kW (4.0 hp) diesel engine (Fig. 4.60). The machine transplants 8 rows with 238 mm row spacing in single pass. Distance between hills is being adjustable from 120 to 140 mm. The drive wheel receives power from the engine through V-belt, cone clutch and gearbox. A propeller shaft from the gearbox gives drive to the transplanting device mounted over the float. The float facilitates the transplanter to slide over the puddle surface. The tray carrying mat type nursery for 8 rows is moved sideways by a mechanism, which converts rotary motion received from the engine through belt-pulley, gear and universal joint shaft into linear motion of a rod connected to the seedling tray having provision to reverse the direction of movement of tray after it reaches the extreme position at one end. Fixed fork with knock out lever type planting fingers (Cranking type) are moved by a four bar linkage to give the designed locus to the tip of the planting finger.

The machine is operated in high/low gear at about half to three fourth throttle depending upon the field condition (Fig. 4.60). For operating the machine in the field, tire ring from front drive wheel and two rear wheels from the float are removed and the field after puddling is left for settlement up to 3 to 4 days depending upon the type of soil. Machine uses mat type seedlings and it can transplant 1.2-1.5 ha/day with the help of 5 persons by working at a speed of 1.1-1.5 km/h. It saves about 65% labour and 40% cost of operation as compared to manual transplanting.

Fig. 4.60: Self-propelled rice transplanter in operation.

Self-propelled walk behind type paddy transplanter

Fig. 4.61: A view of self propelled walk behind type paddy transplanter in operation

Self propelled walk behind type paddy transplanter is a 4 row machine operated by a 3.2 kW petrol engine. The machine consists of power transmission system; handle for steering the machine, main frame and rice transplanting tray, float and two pairs of transplanting units (Fig. 4.61). It has only two lugged wheels and the weight of the machine rests on the lugged wheel and float at the time of transplanting. The same lugged wheels are used for transportation. Power from the engine is transmitted to front traction wheels through gear train and to the transmission housing of transplanting unit through universal shaft. The machine has one forward gear and one reverse gear. Row to row spacing is 30.0 cm and four settings are provided for plant to plant spacing i.e. 12, 14, 18 & 21 cm. Four settings are provided for adjusting the number of hills transplanted/sq. meter. Transplanting depth is also adjustable. It can cover about 0.6 ha/day

Self-propelled 4 wheel type paddy transplanter

Self-propelled 4 wheel type paddy transplanter is a 6 row riding type machine operated by a 12.5 kW petrol engine. The machine consists of power transmission system; handle for steering the machine, main frame and rice transplanting tray, float and two pairs of transplanting units (Fig. 4.62). It has four lugged wheels and the weight of the machine rests on the lugged wheel. The position of rice transplanting tray is adjusted by hydraulic system. The same lugged wheels are used for transportation. The machine has hydrostatic transmission with 5 forward and 5 reverse speeds. Row to row spacing is 30.0 cm and five

Fig. 4.62: A view of self-propelled 4-wheel type paddy transplanter in operation

settings are provided for plant to plant spacing i.e. 12, 14, 16, 18 & 21 cm. Five settings are provided for adjusting the number of hills transplanted/m^2. Transplanting depth is also adjustable.

Transplanter using washed root seedlings

In the design of transplanters, using washed root seedlings, the seedlings having 4-6 leaves and 18-35 cm height are removed from the traditional nurseries just before transplanting and roots are washed with water. In some cases, the roots and tops are pruned and then neatly arranged on the seedling platforms or boxes of the machine for transplanting. In general, it has been found that the labour requirement for uprooting, washing and arranging of seedlings in the transplanter boxes is quite high. It is a 6-row riding type paddy transplanter (Fig. 4.63) using conventional washed root seedlings. It comprises of a transplanting unit, engine, power transmission system, and power driven lugged wheel, machine frame, a wooden float, a seat for the operator and a steering handle. Power from a 4.0 hp petrol engine is given to the transplanting unit as well as the powered wheel through a gearbox. The forward speed can be changed through suitable selection of gears, which would also change the plant to plant spacing. A neutral position exists so that the power to the wheel alone during transportation and the transplanting system can be given independently. Power from the gear box is transmitted to another gear box on the transporting unit through a universal coupling. Gearbox of the transplanting unit has a clutch and a scroll shaft. Scroll shaft moves the nursery boxes in a to and fro motion. It also operates the fingers arm with the help of cams mounted on the shaft and other linkages. The linkages also operate the seedlings pusher for proper nursery picking by the fingers. A steel ring with rubber surface can be fitted on the lugged wheel for machine transportation on hard ground/roads. This can be done very easily in the field itself. Also, two rubber wheels are mounted on the float at its rear end for machine transportation, which are removed during transplanting. Machine is operated and controlled by

Fig. 4.63: 6-row self-propelled rice transplanter using washed root seedlings

the operator sitting on the machine with the help of a steering. Provisions also exist for changing the depth of transplanting, plant to plant spacing and number of plants per hill. Labour required for nursery uprooting for manual transplanting is 60-80 man-hrs/ha. Field capacity of the machine in low gear is 0.12 ha/h and 0.19 ha/h in high gear. Plant to plant spacing in low gear and high gear are kept 15 cm and 21 cm respectively.

Actuating type of fingers: This type of mechanism is used in case of Japanese machines. Each unit has two fingers, which actuate with the help of cams, pinions and push rods. During each stroke, fingers close after picking the seedlings and open at the time of transplanting. The system is very precise but complicated.

Fixed type of fingers: This type of system is common in case of IRRI and Chinese type of machines. In this case number of pickers as desired are mounted on a single bar. Type of pickers in different machines varies but the opening in all the pickers is about 6 mm. The system is very simple and cheap as single bar is given drive in this case. Performance of this system is satisfactory.

Tray moving mechanism

Chain type: This system is cheap and simple and is used in IRRI design. It consists of a bicycle chain, free wheel sprocket and linkages. However, the main disadvantage of this system is only that during reverse motion of the tray at the ends, there is an empty stroke and the tray does not move laterally.

Scroll type: The system is very good except that it is costly and bit complicated. The system works in a closed housing. The scroll shaft is given a drive, which rotates the tray connecting rod laterally through a clamp and a key.

Manually operated low land rice seeder

Transplanting of rice seedlings in puddled soil is a widely accepted cultivation practice. Energy consumed in nursery raising, puddling and transplanting constitutes about 35-40 per cent of the total energy need in cultivating rice. During peak seasons labour is scarce which leads to delay in transplanting. Aged seedlings result in reduced productivity. To overcome the scarcity of labour for transplanting in peak seasons there has been continuous demand for rice transplanter and other direct rice sowing machines. Since rice transplanters require special skills in the preparation of mat type nursery, development of simple equipment for direct sowing of rice is imminent (Singh and Pandey, 2008; Anonymous, 2008 and 2010). The device consists of a handle, drums placed at 200 mm spacing and two lugged wheels at ends. It is designed for sowing pre-germinated rice in puddled field. It sows 2, 4, 6 and 8 rows of paddy with 200 mm row-to-row spacing (Fig. 4.64). A lugged wheel is also provided centrally to drive an agitator in the drums to facilitate easy flow of pre-germinated seeds. The machine floats on two skids. Pre-germinated rice seeds are kept in drums, which have peripheral openings at two ends for seed discharge. The seed rate can be adjusted by adjusting the opening in the cylinder. The field capacity varies from 0.10 to 0.14 ha/h with 63 per cent field efficiency and labour requirement of 15-20 man-h/ha. The traditional method requires a total of 300-320 man-h/ha. Average seed rate with the seeder in the field is 50 kg/ha. Since the crop is row planted, intercultural operations like weeding become easier. The drum-seeded crop comes to harvest at least 7 days earlier as compared to transplanted crop. Number of panicles per plant and number of plants per square metre area are more in the drum seeded crop. Grain yield increased by 6-7 per cent in the drum-seeded crop. Weeds are to be managed with due care and mechanical weeder can be used. There should not be rain within 5-6 days after sowing by the seeder.

a) Four row b) Eight row machine

Fig. 4.64: Manually operated low land paddy seeder

Vegetable Transplanters

Vegetable transplanting is very labour intensive activity. It requires 15-20 man-days per ha for transplanting. The mechanization of vegetable cultivation has been considered as one solution to these problems. Up to now, several kinds of machine and equipment have been developed. Consequently, most of the work is now partly mechanized, and the number of required work-hours has decreased. However, there is still a need for more mechanization. Most growers desire more labour-saving machines and facilities (Anonymous, 2010; Anonymous, 2012; Singh and Pandey, 2008). Vegetable transplanters are thus very important if labour requirement for transplanting is to be reduced. A vegetable transplanting machine must handle a seedling with or without soil, place it in a vertical position and press the soil around its roots and simultaneously release the seedling. In addition to above some transplanting machines also apply water to freshly transplanted seedling. Water is only sprinkled around the seedling by opening a valve when nozzle is above the just transplanted seedling. To achieve vertical position of the seedling relative velocity of the seedling with ground should be zero. Rotating or moving the planting tool in the direction opposite to advance of the machine obtains this. Peripheral speed of the planting disc should be equal to forward speed of the transplanter. Furrow openers in the transplanters generally open a near rectangular furrow that gives best results. Transplanting permits a higher level of stabilization of vegetable harvests as compared to direct seeding. Using transplanters shortens the time that crops are in the field. For second crops, transplanting may allow growers more time for land preparation. This may be very advantage where growers have a limited time to produce for a given market. Raising and transplanting of seedlings, however, requires much labour and cost. Several kinds of machines and equipment have thus been developed in developed countries where labour availability is a problem. A simple transplanting aid pulled through the field and used to carry the plant reduces these labour requirements by about 20 per cent.

There are also various kinds of seedling, which can be used with a mechanical transplanter, including bare seedlings, and seedlings with soil, which are cell plug, paper pot and soil block seedlings. Recently, walking or riding type, semi or fully automatic mechanical transplanter, some equipped with devices for feeding the seedlings, are available. These machines enable high speed and efficiency in planting work. Semi and fully automatic transplanters for leaf vegetables are also available (Table 4.2).

Table 4.2: Classification of vegetable transplanter and appropriate seedling types

Transplanter		Seedlings	
		Bare seedling	
Semi automatic	Walking type Riding type	Seedling with soil	Soil block seedling Cell plug seedling Paper pot seedling
Fully automatic	Walking type Riding type	Seedling with soil	Cell plug seedling Pulp (paper) mold cell pot seedling Linked paper pot seedling

Semi-automatic transplanter

Semi automatic transplanters consist of the transplanting mechanism, manual-feeding mechanism, drives system, etc. Rate of planting is limited by the speed of one by one-manual seedling feeding. Generally, it takes about two second to feed a seedling by hand, and this work is very tedious in a large field. This type of transplanter, however, has some advantages, such as it is simple mechanism, low price, and wide adaptability to various seedling types. Semi automatic transplanters are thus widely used in many countries. The transplanting section consists of a furrow opener, pair of press wheels, and two disks for planting; one of the discs is covered with foam or rubber. From the seat the operator feeds the seedling by hand to the disks, which catch the seedling and automatically place it at bottom of furrow. Press wheels then firm up the soil. Rate of planting is about 2,000 seedlings per hour, depending on the operator's skill. Rate of work is three to five hours per ha depending on plant spacing. To keep the seedling vertical, the discs must release it when it is vertically below the axis of the discs. A premature release results in backward bending of the seedling because it is just held planting when released. Delayed release results in forward bending because horizontal component of peripheral speed of planting disc starts reducing and seedling gets some relative motion w.r.t. soil.

Tractor operated vegetable transplanter (Picker wheel type)

A two row semi-automatic planter having picker wheel type metering mechanism has been developed to transplant the wash root type seedlings on the beds as well as on the flat fields. The machine consists of a frame, four lugged ground wheels, three seedling trays, and two seats for the operators, furrow opener, compaction wheels, finger guide funnel, picker wheel type metering mechanism and a water tanker (Fig. 4.65). The picking forks have spring mounted rubber flappers, which open before passing through the tunnel and close during its passage. Again the flappers open at the bottom end of the tunnel to release the seedlings in a furrow. The wheels compact the soil around the seedlings. The power from the wheel is supplied to the planting mechanism through shaft, chain and sprockets. The plant spacing in the machine is kept at 300 mm but it can be varied by changing the sprockets or number of fingers. Two persons (one for each row) are required to place the seedlings in the flappers when these open at the top position. To increase the planting speed, number of persons required to feed the seedlings is increased to two per row. After the seedling is dropped in the furrow, the soil is compacted around it with the help of two moving inclined wheels. Average field capacity of machine is 0.1 ha/h at forward speed of 1 km/h.

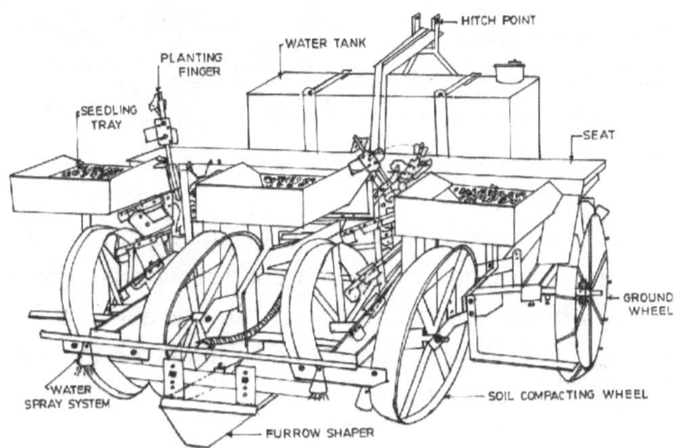

a) Details of vegetable transplanter

b) Vegetable transplanter in use

Fig. 4.65: Tractor-operated two-row vegetable transplanter (Picker wheel type) (*Source:* Anonymous, 2010)

Tractor operated vegetable transplanter (Plug type)

Tractor operated plug type vegetable transplanter (Fig. 4.66) consisted of main frame with hitching system, ground wheel, shoe type furrow openers, compaction wheel, operator's seats, two depth control wheels and plug type metering mechanism (Anonymous 2010). It employs press wheels inclined at an angle of 15° with the vertical as soil covering device. The plant spacing can be adjusted by changing sprockets / pulleys or by changing in alternate holes. The power is transmitted from ground wheel (750 mm diameter) to gear box through chain and sprockets. For transmitting power from gear box to discs, belt and pulley are employed. In transplanting three discs of 460 mm diameter are used. The power is reduced at a ratio of 2:1 from gear box to discs. The auxiliary frames are provided to mount furrow openers, furrow closures, operator's seat and two depth control wheels. The impact of the seedling with soil block helps in its placement. The average field capacity of transplanter is 0.14 ha/h for chilli, tomato and brinjal with 450 mm row spacing. The working width of machine is 1350 mm.

Fig. 4.66: Tractor operated plug type vegetable transplanter

Fully automatic tractor operated vegetable transplanter

Fully automatic vegetable transplanters consist of the transplanting mechanism, automatic seedling feeding mechanism, drive system, etc (Fig. 4.67). Transplanting is fully automatic, and rate of planting is thus high. The operator only needs to steer the machine and replace the seedling trays (Anonymous 2013). However, this type of transplanter can utilize only an appropriate seedling type

Fig. 4.67: Fully automatic tractor operated vegetable transplanter

and seedling arrangement on tray. Naturally, some transplanters synchronize burning or cutting holes through the mulch plastic for full-bed plastic mulch culture, which is now popular for lettuce. Planting depth is controlled automatically. This transplanter is a row type with four-cycle gasoline 1.67 kW engine. A flexible plastic cell tray (the under-tray is the same as that used for rice) is employed. The cells are 30 mm square with 128 cells (8x16) per tray, and 25 mm square with 200 cells (10x20) per tray. Cell plugs are shaped like an inverted pyramid. The fingers, which operate almost like chopsticks, pick up a cell plug from the tray and throw it in the bill type opener, which plants the plug after pre-pressing the soil surface with wheels. Finally conical wheels press the soil around both sides of the plug. Hill spacing must be between 28-56 cm, and suitable ridge height is 0-25 cm. Rate of planting is about one seedling per second, and rate of work for a single operator is 2 to 2.5 hour per ha depending on hill spacing. An automatic transplanter for pulp mold cell pot seedling is also available. This transplanter is a row type with four-cycle gasoline engine. The seedling pot is made of molded paper, and the under-tray is the same as that used for rice. Paper pots are available in 30 mm square (144 pots (8x16) per tray); 23 mm square 200 pots (10x20) per tray size and other sizes. The pot shape is pyramidal. A planting finger scratches and holds a seedling pot, then plants it in a hole using a dig-arm. Press wheels then press the soil around both sides of the seedling. The planting finger consists of two blades-like plate and planting fork. The blades hold the seedling and planting fork places it in the soil. Hill spacing must be between 22-50 cm, and suitable ridge height is 0-25 cm. Rate of work for a single operator is 1.7 to 3 hour per ha depending on hill spacing. Another automatic transplanter for chained pulp pot seedlings is a row type with four-cycle gasoline 1.7 kW engine. The chained paper pots are in a row of 264 pots. The paper pot is 30 mm in diameter and 5 cm high. Shape is cylindrical. A rotating type planting finger picks a paper pot, then guides it through the slider and into the soil. The planting finger splits the paper pot one by one before transplanting. Conical press wheels press soil against both sides of the seedling. Hill spacing must be between 24-64 cm, and suitable ridge height is 0-25 cm. Rate of planting is about two seedlings per second, and rate of work for a single operator is 1.5 to 2.4 hour per ha depending on hill spacing.

Tractor operated revolving magazine type vegetable transplanter

A two row vegetable transplanter with revolving magazine type metering mechanism has been developed by PAU Ludhiana (Fig. 4.68). The machine consisted of furrow opener shoe type, packing wheel, main axle, frame, three point hitch system, ground wheels, seedling trays and power transmission (Anonymous, 2013). The average labour requirement for transplanting is 32.76 man-h/ha for brinjal and for tomato 35.08 man-h/ha as compared to 254 man-h/ha in manual transplanting. The average field capacity of the machine was 0.12-0.15 ha/h for different vegetable crops.

Fig. 4.68: Tractor operated two row revolving magazine type vegetable transplanter

References

Agrawal K.N., Ghadge S.V., Singh RKP., Satapathy K.K., Pandey M.M. 2003. Directory of Horticultural Tools and Machinery for Hill Region. AICRP on FIM, ICAR Research Complex for NEH Region.

Anonymous. 2008. Research Highlight. AICRP on Farm Implements and Machinery, CIAE Bhopal. *Technical Bulletin No.*: CIAE/2008/141.

Anonymous. 2008a. Success Stories. AICRP on Farm Implements and Machinery, CIAE Bhopal. *Extension Bulletin No.*: CIAE/FIM/2008/80.

Anonymous. 2010. Research Highlight. AICRP on Farm Implements and Machinery, CIAE Bhopal. *Technical Bulletin No.*: CIAE/2010/151.

Anonymous. 2012. Directory of Successful Farm Machinery in SAARC Countries. SAARC Agriculture Centre. BARC Complex, Farmgate, Dhaka – 1215 (Bangladesh).

Anonymous. 2013. Research Highlight. AICRP on Farm Implements and Machinery, CIAE Bhopal. *Technical Bulletin No.*: CIAE/2013/158.

Anonymous. 2013a. Success Stories. AICRP on Farm Implements and Machinery, CIAE Bhopal. *Extension Bulletin No.*: CIAE/FIM/2013/40.

Bhardwaj K C; Pandey M M; Singh Gyanendra. 2004. Horticultural Tools and Equipment. CIAE Bhopal. *Technical Bulletin No.*: CIAE/2004/110.

Garg I K. 1992. Investigations on the design and operational aspects of manually pulled engine operated paddy transplanter using mat type seedlings, unpublished Ph.D. Thesis. PAU Ludhiana.

Garg I K. Singh C P and Sharma V K. 1982. Influence of selected seedling mat parameters and planting speed on performance of Rice Transplanter. *AMA*, 13(2):27-33.

Garg I K; Sharma V K. 1985. Design, Development and Evaluation of PAU Riding type Engine Operated Paddy Transplanter Using MAT type Seedlings. *Proc. ISAE. SJC.* 1(2):57-63

Garg I K; Sharma V K. 1987. Riding type engine operated paddy transplanter. *Invention Intelligence.* 22 (9-10): 368-74.

Garg I K. 1987. Annual Report of ICAR Co-ordinated scheme for Research and Development of Farm Implements and Machinery. Department of Farm Power and Machinery, PAU, Ludhiana. Pp 47-69.

Garg I K; Sharma V K. 1984. Development and evaluation of a manually operated paddy transplanter. *J. Agric. Engg.* ISAE XXI (1-2) 17-24

Garg I K; Singh Surendra. 2002. Farm equipment for Punjab agriculture. Department of Farm Power & Machinery, Punjab Agricultural University, Ludhiana.

Pandey M M; Ganesan S. 2005. Farm Mechanization Package for Dryland Agriculture, Technical Bulletin No: CIAE/FIM/2005/117. Central Institute of Agricultural Engineering, Bhopal.

Pandey M M; Majumdar K L; Singh Gyanendra; Singh Gajendra. 1997. Farm Machinery Research Digest. Technical Bulletin No. CIAE/97/69, Central Institute of Agricultural Engineering, Bhopal, 328 p.

Pandey M M; S Ganesan; R K Tiwari. 2006. Improved Farm Tools and Equipment for North Eastern Hills Region, Technical Bulletin No. CIAE/2006/121. Central Institute of Agricultural Engineering, Bhopal.

Shukla L N; Tandon S K; Verma S R. 1984. Development and field evaluation of coulter attachment for direct drilling. Published in AMA, Vol. XV(3), 19-21, 24, Summer 1984, Tokyo, Japan.

Shukla L N; Chauhan A M; Dhaliwal I S. 1990. Minimum tillage machine for developing countries for early 21st century. Proceddings of the International Agricultural Engineering Conference and Exhibition, Bangkok, Thailand, 1: 381-388, Dec. 3-6, 1990.

Shukla L N; Chauhan A M; Dhaliwal I S; Verma S R. 1996. Development of Minimum Till Planting Machinery. *AMA*, Vol. 27(4):15-18

Shukla L N; Verma S R; Singh Amar. 1978. Development of a semi-automatic tractor drawn roto drum sugarcane planter. *J.Agric. Engg.* XV(4) : 211-214.

Singh A K; Sharma M P. 2008. IISR sugarcane cutter - planter. Operation Manual No. AE/08/01, IISR Lucknow: p 6.

Singh Surendra. 2007. Farm Machinery – Principles and Applications. Directorate of Information & Publication of Agriculture, Indian Council of Agricultural Research, Krishi Anusandhan Bhawan-I, Pusa Campus, New Delhi.

Singh Surendra; Pandey M M. 2008. X Plan Achievements (2002-2007). AICRP on Farm Implements and Machinery, CIAE Bhopal. Technical Bulletin No.: CIAE/2008/137.

Singh Surendra; Verma S R. 2009. Farm Machinery Maintenance & Management. Directorate of Information & Publication of Agriculture, Indian Council of Agricultural Research, Krishi Anusandhan Bhawan-I, Pusa Campus, New Delhi.

Singh Surendra; Sharma V K; Gupta P K; Chauhan A M. 1985. A ridge planter for winter maize. *J. of Agril. Engg.* 22(2): 115-18.

Singh Surendra; Sharma V K; Gupta P K; Chauhan A M. 1988. Design, development and evaluation of two-row tractor mounted ridge planter for winter maize. *Indian J. of Agricultural Sciences.* 58(1): 45-47.

Singh P R. 2014. Souvenir, Tractor and Agricultural Machinery Manufacturers' Meet (TAMM 2014). Held at IISR Lucknow during February 8-9.

Srivastava N S L; Sharma M P. 1977. A power operated sett cutting machine. *Journal of Agril. Engg.* 14(3):119-120.

Verma S R; Sharma V K; Singh Surendra; Ahuja S S; Singh Santokh (eds.). 1992. Proceedings of National Colloquium on Potato Mechanisation in India. *Punjab Agricultural University, Ludhiana.*

Yadav RNS. 2003. Sugarcane Production Machinery, CIAE, Bhopal. Bulletin No. CIAE/NATP/-SM/2002/95.

5 Interculture and Weeding Equipment

Weeds are a serious menace and these reduce the quality and yield of crops substantially. Chemical as well as mechanical methods are being used to control the weeds. The weeding was traditionally done by "khurpa" up to mid 60s. This tool requires its operation in a sitting posture, which reduces the work rate. The use of long handled tools like "kasola" became popular because these allowed their operation in a standing posture, which increases the mobility and thus the work rate. Weeder is a mechanical device to remove the weeds (unwanted plants) from the cultivated crop. A wide variety of weeders are used in the countries, which may be categorized in three groups viz. manual weeder, animal or tractor drawn weeder and power weeder or self-propelled rotary weeder (Singh, 2007; Pandey *et al.*, 1997; Pandey and Ganesan, 2005).

Hand weeder

Hand weeders are operated by a man/women worker to remove the weed between the crop rows or between the plants. Mostly hand weeder is design to operate in the row by a single man. Weeder mulcher is used for breaking soil crust over germinating seeds. This also controls and destroys young weeds just after field-crop has begun to grow. Weeders with coil-spring teeth are excellent tools for making mulch. Peg type hoe consists of a handle, adjustable V-blade and peg holding roller. The top soil crust is broken by roller while the weeds uprooting and soil mulching are done by the blade through push and pull action of the unit. It is used for weeding in groundnut, soybean, sunflower, and sorghum crops.

Grubber (Three tined hand hoe)

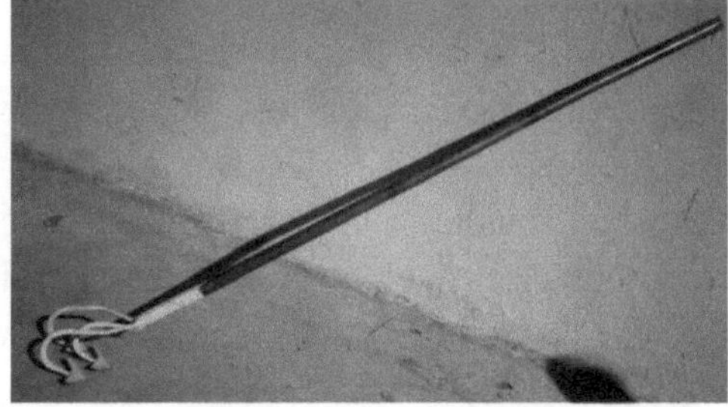

Fig. 5.1: Three tined hand hoe

It is a simple and light weight, manually operated equipment for weeding and interculture in upland row crops (Pandey and Ganesan, 2005). It consists of a long handle, ferrule, three tynes and sweep type blades (Fig. 5.1). The operator uses pull force to break the soil crust and uproot the weeds. The three tined hand hoe is one of the widely used hoes for weeding and interculture in horticultural crops. It is a long handled weeding tool and can be easily maneuvered between the rows and around the plants. The tool essentially consists of three curved tines, which are bent to appropriate shapes. The working blade tips are forged, flattened and sharpened. The section of these tines is round or square depending upon the design. The tines are made from medium carbon steel and the tips are hardened to 40-45 HRC. The tines are either welded to the ferrule or secured in the ferrule by I-bolt, which helps in adding the number of tines. The ferrule is made from mild steel in which the wooden handle is inserted. The handle is made from good quality wood or bamboo. For operation the tool is held in standing position and pulled towards the operator, which causes suction at the tips/ edges of the tines and penetrate into the soil. The pulling action uproots the weeds and also aerates the soil. In some of the designs, small V-shape blades are welded to the tips.

Twin wheel hoe

Twin wheel hoe consists of twin wheels, frame, V-blade with tyne, U-clamp, scraper and handle (Fig. 5.2), Pandey and Ganesan (2005). The cutting and uprooting of weeds in field is done through push and pull type action of the equipment. The equipment is operated at optimum condition of field for better output. Women workers likes the equipment.

Fig. 5.2: Twin wheel hoe

Straight blade hand hoe

The straight blade hand hoe is a long handled hand tool for operation between the crop rows. It consists of a blade, curved arm, ferrule and a long wooden handle (Fig. 5.3), Pandey and Ganesan (2005). The curved arm joins the blade with the ferrule to which the handle is fixed. The blade, which performs the cutting/ uprooting of the weeds, besides stirring the soil is made from medium carbon steel hardened to 350-450 HB. The curved arm and the ferrule are made from mild steel. The handle is made from good quality wood or bamboo. Being a long handled tool, the straight blade hand hoe is operated in the standing posture by pulling action. The pulling action of the blade into the soil cuts or uproots the weeds in between the rows of the crop. The cut or uprooted weeds are buried under the soil and thus creates mulch. The straight blade hand hoes are available in various sizes. The hoe is used for weeding of the vegetable crops planted in rows and nursery bed preparation.

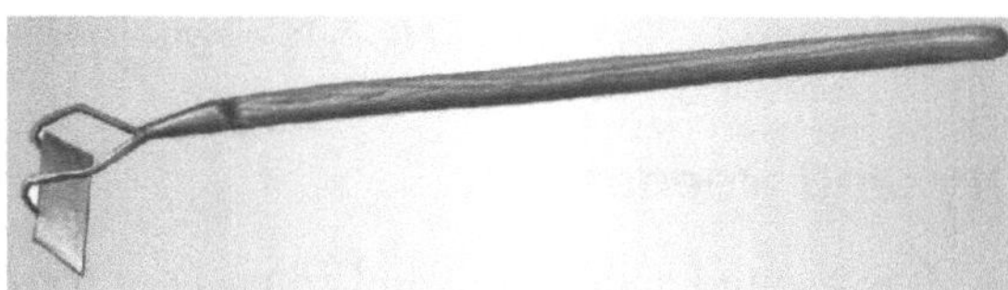

Fig. 5.3: Straight blade hand hoe

V-blade hand hoe

The V-blade hand hoe is a long handled weeding tool for operation in between the crop rows. The hoe consists of a V-blade, arms, ferrule and a wooden handle (Fig. 5.4), Pandey and Ganesan (2005). The arms are welded to the ferrule. Wooden handle is inserted in the ferrule. The blade of the hoe is the important component, which enters into the soil and performs the cutting and uprooting of weeds. The blade is fabricated from medium carbon steel and the edges of the blade are hardened to 40-45 HRC. The arms and the ferrule are made from mild steel. The handle is made from high quality wood or bamboo. Being a long handled tool, the hoe is operated in the standing posture by pulling action towards the operator. The pulling actions cause penetration of the blade into the soil and cut or uproot the weeds. Because of the V-shape, the blade creates small furrows between the crop rows and also earthing of the plants. This operation facilitates the flow of irrigation water to the root zone of the plant. The hoe is used for weeding of the vegetable crop planted in the rows and earthing operation.

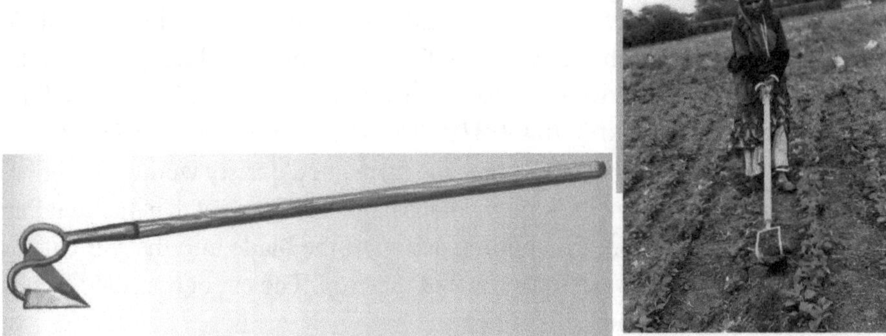

Fig. 5.4: V-blade hand hoe

Dry land weeder

It is a long handled tool and consists of rotor with diamonds shaped pegs, cutting blade, frame and handle (Fig. 5.5), Pandey and Ganesan (2005). The rotor consists of two circular discs, mild steel bars, and diamonds shaped pegs welded on the round bars. The pegs are arranged in staggered fashion. An axle is provided in the rotor and joined to the frame. The blade has two arms for adjusting depth of penetration in the soil. The cutting edge of the blade is slightly convex and made sharp. The frame is welded to the handle. The weeder is operated by imparting a push pull action. The pegs of the rotor penetrate in to the soil and create soil mulch and the blades cut and uproot weeds. The weeder is operated between crop rows. Its capacity is 0.05 ha/day. It is used for removal of shallow rooted weeds in row crops. It has been designed ergonomically for easy operation. Useful in dry land and garden land crops and is ideal at a soil moisture content of 8 to 10 per cent.

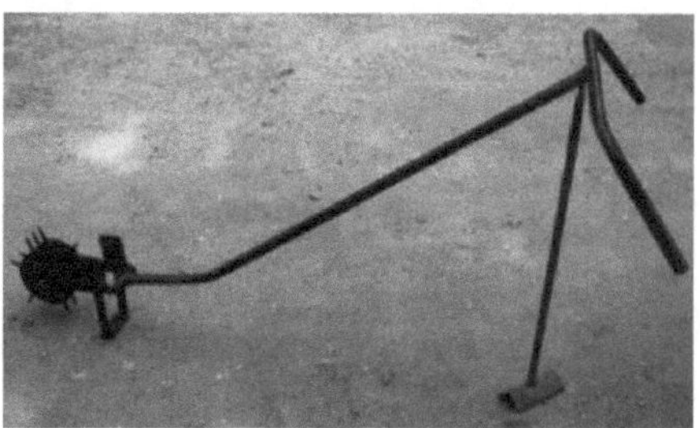

Fig. 5.5: Dry land weeder

Dry land weeder (Star type)

The dry land weeder (star type) is suitable for operation in light soils for interculture. It consists of the three star wheels mounted on the gang, a curved blade, frame and handle assembly (Fig. 5.6), Pandey and Ganesan (2005). The star wheels and the blade assembly are mounted on the frame to which handle assembly is joined. The star wheels and the blade are important component of the weeder which are made from medium carbon steel and the edges are hardened to 40-45 HRC. In some models, the star wheels are made from mild steel. The frame assembly is fabricated from structural mild steel and handle assembly from thin walled mild steel pipes. The handle height is adjustable to suit the operator's working height. For operation the weeder is repeatedly pulled and pushed between the rows of crop. The star wheels break the soil crust and blade following it, cuts and uproots the weeds. The blade can be adjusted to vary the depth of working. Dry land weeder (star type) is used for breaking the soil crust, aeration of soil, cutting and uprooting of weeds in vegetable gardens, fruit tree basins, grape vine yard etc.

Fig. 5.6: Dry land weeder (Star type)

Dry land weeder (Peg type)

The dry land weeder (peg type) is one of the useful manually operated weeders for operation between the crop rows. It consists of a roller, which has two mild steel discs joined by mild steel rods (Fig. 5.7), Pandey and Ganesan

(2005). The axle passes through the centre of discs and is mounted on the two arms, which also constitutes the frame. The small diamonds shaped pegs are welded on the rods in a staggered fashion. The complete roller assembly is made of mild steel. The V-shaped blade follows the roller assembly and is mounted on the arms. The blade is fabricated from medium carbon steel and forged to shape. The cutting edges are hardened to 40-45 HRC. The height of the blade can be adjusted according to the working depth. The arms are joined to the handle assembly, which is made from thin walled pipes. The height of the handle can also be adjusted according to the operator. For operation the weeder is repeatedly pushed and pulled in between the crop rows in the standing position. The diamonds shaped pegs penetrate into the soil and the rolling action pulverize the soil. The blade in the push mode penetrates into the soil and cuts or uproots the weeds. The weeder is used for removing weeds in vegetable gardens, basins of orchard trees and vineyard plantations. It is also used for breaking the soil crust and creation of soil mulch.

Fig. 5.7: Dry land weeder (Peg type)

Hand cultivator

The hand cultivator is a useful hand tool, which is operated in a squatting position to create soil mulch and uprooting weeds (Pandey *et al.*, 1997; Pandey and Ganesan, 2005). The design of the hand cultivator varies with the region having different soil types. The tool essentially consists of prongs, which are bent at the working end, gathered at the holding end and joined at the ferrule to which a wooden handle is attached (Fig. 5.8). The number of prongs varies from 3 to 7 depending upon the design. The working tips of the prongs are usually forged to diamond point for easier penetration into the soil. These prongs are fabricated from medium carbon steel rods and the working tips are hardened to 370- 470 HB. A semicircular strip is provided near the gathering end to maintain the distance between the prongs and also keep them diverged at the working end. For operation, the tool is pulled towards the operator, which manipulates the soil, uprooting small weeds. In some models, the numbers of prongs are reduced to three and the small sweeps are welded to the tip. In these models, a long handle is provided which facilitates operation in a standing position. The tool can be manoeuvred easily in the crop rows and around the plant for effective removal of weeds. The hand cultivators with short handle are available in various design and sizes and have 3, 5 or 7 prongs. The tool is used for weeding, creating soil mulch, aeration and moisture conservation and breaking of the soil crust in horticultural crops.

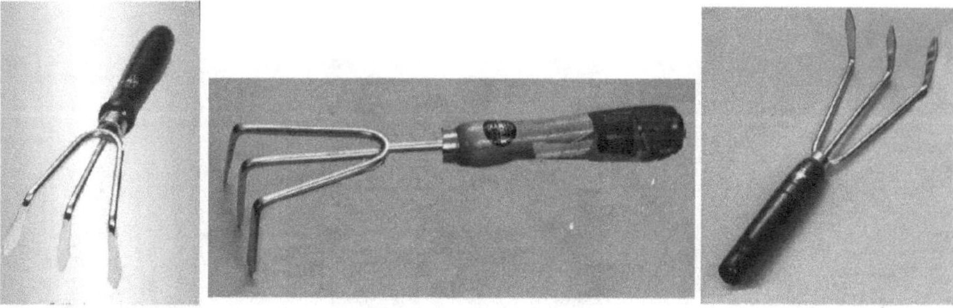

Fig. 5.8: Hand cultivator

Wheel hand hoe

It is manually operated push-pull type weeder (Anonymous, 2010; Garg and Singh, 2002, Agrawal *et al.*, 2003). The wheel hoe is a widely accepted weeding tool for weeding and interculture in row crops (Fig. 5.9). Details of wheel hand hoe are given in (Fig. 5.10). It is a long handled tools operated by push and pull action. The number of wheel varies from one to two and the diameter depends upon the design. The frame has got a provision to accommodate different types of soil working tools such as straight blade, reversible blades, sweeps, V-blade, tine cultivator, pronged hoe, miniature furrower, spike harrow (rake) etc which can be operated by a single person. The handle assembly has a provision to adjust the height of the handle to suit the operator. All the soil working components of the tool are made from medium carbon steel and hardened to 40-45 HRC. The working depth of the tool can be adjusted with the help of clamp or through the plate with multiple holes provided in the frame and welded to the tool assembly. The handle height is also adjustable. Some of the designs are provided with a pulling ring in the frame, which enables it to be operated by two persons. For operation, the working depth of the tool and handle height is adjusted and the wheel hoe is operated by repeated push- pull action which allows the soil working components to penetrate into the soil and cut/uproot the weeds in between the crop rows. With this action, the weeds also get buried in the soil. It covers 0.1 ha/day. It is manually operated walking type device and it can be used for various farm operations right from opening furrows for sowing seeds to intercultural and earthing up operations depending upon the type of attachment used. It requires more than 20 cm row-to-row spacing for easy operation. It saves 70-75% on labour and operating time and 80% on cost of operation and also results in 5-8% increase in yield compared to conventional method of weeding using Khurpi.

Fig. 5.9: Wheel hand hoe with one tyne and three tynes

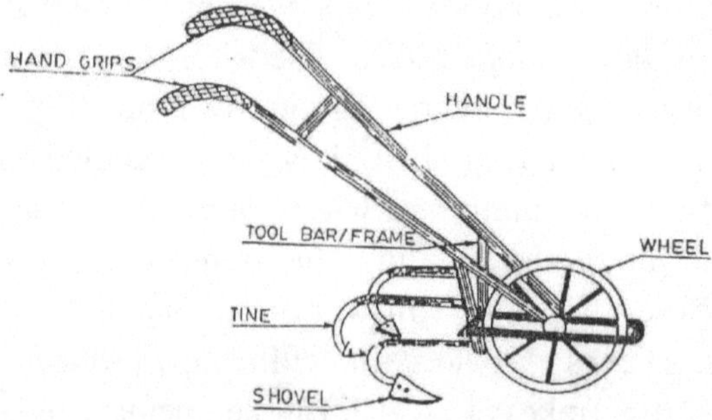

Fig. 5.10: Details of wheel hand hoe

Finger type rotary weeder

The two row paddy weeder consists of two finger type rotary weeder, handle and frame (Fig. 5.11), Anonymous, 2010. The saving in cost is 54% and saving in time is 80% with this weeder as compared to traditional method. Comparing easiness in operation of single, double and three row weeder the farmers preferre single row weeder which is gender friendly also. The two row and three row rotary weeder is recommended for Transfer of Technology.

Fig. 5.11: Finger type rotary weeder

Cono weeder

Traditionally hand weeding was popular but due to increased wages and scarcity of labourer, the farmers are facing problems and needs suitable mechanical weeders for low land rice fields. Weeding in the paddy fields is generally performed in the standing water or muddy field conditions. Some of the manually operated mechanical weeders such as Cono weeder, Japani paddy weeder and Ambika paddy weeder are commercially available. The Cono weeder and Japani weeders are comparatively heavier and having two rolls. Whereas Ambika paddy weeder, developed at IGKV, Raipur, is having only one roller and is very light in weight (weight 3.5 kg) and cheaper and reported more effective and efficient. The Ambika paddy weeder can be safely operated by one male or female worker within the 20 cm spacing of rice crop in a push-pull mode. The working efficiency of a worker in weeding is increased by 8-10 times than hand weeding. The cono weeder consists of two rotors, float, frame and handles (Fig. 5.12), Anonymous, 2008, 2010, 2012. The rotors are cone frustum in shape, smooth and serrated strips are welded on the surface along its length. The rotors are mounted in tandem with opposite orientation. The float, rotors and handle are joined to the frame. The float controls working depth and does not allow rotor assembly to sink in the puddle. The cono weeder is operated by pushing action. The orientation of rotors create a back and forth movement in the top 3 cm of soil. The cono weeder is used to remove weeds between rows of paddy crop efficiently. It is easy to operate, and does not sink in the puddle. Push pull operation of cono weeder in between rows makes weeding effective. Its capacity is 0.18 ha/day.

Fig. 5.12: Cono weeder

Animal drawn weeder

The faster intercultivation with low soil compaction during the limited available time results into higher crop growth and yield due to aeration at root zone. Animal drawn weeder (Blade hoe) are useful for faster weeding and intercultivation in soybean and other wide spaced crops compared to the manual weeding by hand hoe (Fig. 5.13), Pandey *et al.*, 1997. It can be operated by a pair of bullocks and can cover two rows at a time. It can cover about 0.1-0.15 ha/h.

Fig. 5.13: Animal drawn weeder

Animal drawn sweeps

It is an animal drawn implement having facilities for adjustment of row-to-row spacing and depth. It is used as an intercultural implement in row crops. It consists of duck-foot type sweeps clamped on a common tool bar and a pair of wheels for depth adjustment (Fig. 5.14), Pandey *et al.*, 1997. It removes the shallow rooted weeds and provides a soil mulch so as to conserve the soil moisture from evaporation loss. It saves 98 per cent labour and operating time and 75 per cent on cost of operation compared to conventional method of manual weeding using hand hoes. Row spacing is adjustable. Field capacity is 0.2 ha/h and weeding efficiency 91 per cent. It is used as an intercultural implement in row crops.

Fig. 5.14: Animal drawn sweeps

Self-propelled light weight power weeder

Power tillers are commercially successful only in areas for wetland cultivation. The use of power tillers in upland cultivation is negligible. In paddy cultivation especially on small and medium farms and for the farm operations in hilly areas, orchards and forestry, power tiller is the only solution. The existing power tillers are too heavy to be used for hilly area. The equipment has to be lifted for shifting from one field to another. So there was a need for developing a lightweight power tiller, which can be used for doing various farm operations. Also, cost of this power tiller should not be very high. It should be able to help small farmers to mechanize the farm operations at lower initial cost. The operational cost of this power tiller for doing various jobs should be less in comparison to traditional power system being used presently. It would also reduce the labour requirement and would enhance efficiency and effectiveness of operation. The development of machine would help to mechanize hilly area, small land holding, and plantation crops and also wide row crops like cotton, sunflower, sugarcane etc. Also, development of this machine will help to mechanize weeding operation under the horticultural trees, which at present is a big problem. In hilly area the cultivation of land is generally done by using bullock-operated implements and intercultural operation is carried out manually. Some farmers used tractor with cultivator or harrow for interculture operation in orchards. This method is not only costly but also in effective due to non-weeding near the plant row due to obstruction of branches of the plants.

A light weight self-propelled power weeder consists of a 4.1 kW diesel engine mounted on the power tiller chassis, power transmission system, two MS wheels, a frame and a rotary (Fig. 5.15), Anonymous, 2010, 2012; Singh and Pandey, 2008. The power from the engine is transmitted with the help of belt and chain to the rotary and through gear train to the ground wheels. The rotary has been provided with 16 blades fitted on high-pressure pipe of 37.5 mm diameter with the help of nuts and bolts to the flanges. For depth adjustment two skids made of flat are provided on both sides of power tiller. A power cut off device is provided to engage or disengage the power to the rotary system. The rotary blades are made of high carbon steel (EN-31), which acts as working tool for weeding or seedbed preparation. The working width of the machine is about 45 cm (adjustable). The wheels with lugs are provided for traction. The unit is employed with simple mechanism of power transmission as the gear train is replaced with the help of pulley and sprockets, resulting in less cost of maintenance. The speed of power weeder can be varied from 2.3-2.5 km/h with effective working width of 550 mm giving field capacity of 0.10 to 0.13 ha/h. The equipment saves 90% operating time and 30% in cost of weeding as compared to hand weeding by Khurpi.

Power tiller operated sweep cultivator

a) Light weight power weeder in use

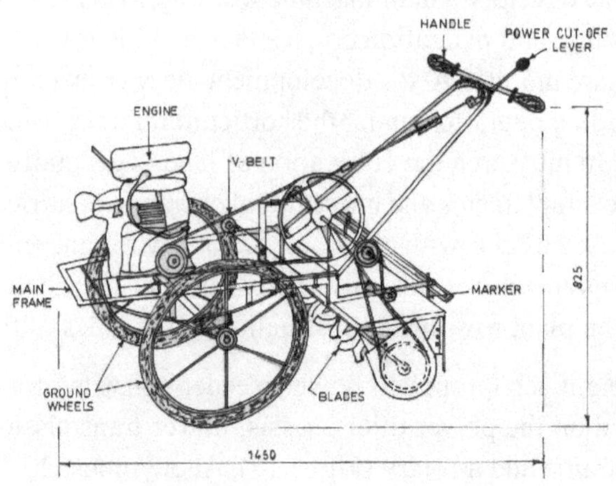

b) Details of power weeder

Fig. 5.15: Self-propelled light weight power weeder

This machine has been specially designed for operation with a power tiller of 6-8 hp to perform interculture operations in standing crop in soybean, sorghum, Bengal gram, pigeon pea etc where the row spacing is wide enough for the power tiller to pass without damage to the plants (Fig. 5.16), Pandey *et al.*, 1997. One depth control wheel is provided at its rear for adjusting the depth of operation. It is suitable for medium and light soils. It is provided with depth control gauge wheels. Some of its major components are the main frame with hitch system, handle, and drive wheel, tynes. Field capacity is 0.18-0.25 ha/h and labour requirement 4-6 man-h/ha.

Fig. 5.16: Power tiller operated sweep cultivator

Power weeder for low land rice

A twin row engine operated weeder having weight of 17 kg and provided with float and rotary cutting blades has been developed with the help of Premier Power Equipments & Product Pvt. Ltd. Coimbatore and TNAU Coimbatore to reduce the labour requirement and drudgery involved in manual weeding in wet land cultivation of rice (Fig. 5.17), Anonymous, 2010. This weeder performs well under all soil conditions. The weeding is done due to the rotary cutting blades. The motorised power weeder consists of 1.5 hp single cylinder, 2-stroke air cooled petrol engine, gear box (worm gear), main frame, rotary wheel, float, handle and controls. Engine speed is 6000-6500 rpm with gear reduction of 1:40. The engine and all

Fig. 5.17: Power weeder for low land rice

other accessories are mounted on main frame made of mild steel pipe. Engine axle is fitted in the final drive mechanism with frictionless roller bearings with packed seal. Throttle is provided on left side of handle to control the speed of engine. Two cutter wheels of diameter 300 mm and width 150 mm served as weeding part. Each cutter wheel has four sharp blades fitted on the periphery. These cutter wheels are fitted on axle and secured. Rotary speed is 150-175 rpm. The machine can cover 0.70 ha/day. Plant damage is less than 1%.

Power weeder for horticultural crops

The power weeder is suitable for interculture operations in orchard crops. It consists of 4 blades, mounted on an adjustable shaft suitable for weeding in various horticulture crops (Fig. 5.18), Anonymous, 2008. The blades are rotated by 1.4 hp Mitsubishi engine operated with petrol engine. The effective with of the weeder is 25 cm. Depending on the requirement; the weeder can be operated on 4, 3 or 2 wheels with 25 cm, 20 cm or 16 cm operating width. The weeder is of 1m height and has 10.4 kg weight. This is suitable for weeding in chilies, turmeric, sugarcane, banana, vegetable crops, orange, lemon, papaya, medicinal plants and other row crops like jowar, groundnut and mulberry.

Fig. 5.18: Power weeder in operation

Tractor operated 3-row rotary weeder

Manual weeding is laborious and time consuming and hence efficient mechanical weeders are being developed to obtain good yields from the farm. A tractor operated rotary weeder has been designed and developed (Anonymous, 2008, 2010, 2012). Rotary weeder consists of a main frame, gearbox, three rotary weeding blade assemblies, a square shaft for transmission of power from gearbox to rotary assemblies and sets of sprockets and chains (Fig. 5.19). A standard 3-point hitch arrangement has been provided to mount the frame to tractor. Power from tractor PTO is transmitted to main square shaft through gearbox mounted on mainframe and set of sprockets and chain. The speed reduction from PTO to gearbox is 5:9. Power from gearbox to blade assemblies is provided by 40 mm square shaft. This facilitates adjustment of row-to-row spacing from approximately 675 to 1165 mm. The weeder encompasses three sets of rotary blade assemblies. Power to these assemblies is provided from main shaft with the help of chain and sprockets. The chain housings provide structural support to blade assemblies. Two flanges are provided on each side of chain housing. Each flange carries four blades. The rotational speed of blade assembly is 1.33 times the rotational speed of main shaft. The machine has been equipped with new heavy-duty gearbox and gear type idlers The machine saves 54% labour and 74% cost of operation as compared to traditional method. The field capacity of machine is 0.24 ha/h with weeding efficiency of 83-87%. Tractor operated rotary weeder are used for weeding and intercultural operations in between row crops like sugarcane, cotton, maize, etc. The multi row rotary weeder consists of a set of cutting blades, which penetrate in to the soil, cause removing the weeds in the crop rows. The power train consists of a gearbox containing a set of bevel gears having transmission ratio 2:1. The driving shaft of gearbox is coupled with PTO shaft of the tractor and driven shaft is coupled with rotary units by means of chain and sprocket power transmission systems. The “L” shaped cutting edge has been sharpened for easy cutting and fixed at an optimum angle of inclination of 50^0 to horizontal. The cutting blade has also been used as an inclined plane for elevating and converging the soil to the rotating blades to perform cutting the weeds and pulverizing the soil.

Fig. 5.19: Tractor operated 3-row rotary weeder

Tractor operated 9 tine weeder

Nine tyne tractor operated cultivators are often used for intercultural operations, till crop height is less than 40 cm. The tynes of cultivator are arranged rather adjusted in such a way that three tynes run in between two rows of crop and three inter row spacing are covered in single run (Fig. 5.20), Anonymous, 2012. It is advisable to use sweep shovels in place of reversible shovels. Sweep shovels completely cover the spacing and no weed is left in the covered space.

Fig. 5.20: Tractor operated 9 tine weeder

References

Agrawal K N; Ghadge S V; Singh R K P; Satapathy K K; Pandey M M. 2003. Directory of Horticultural Tools and Machinery for Hill Region. AICRP on FIM, ICAR Research Complex for NEH Region.

Anonymous. 2008. Research Highlight. AICRP on Farm Implements and Machinery, CIAE Bhopal. Technical Bulletin No.: CIAE/2008/141.

Anonymous. 2010. Research Highlight. AICRP on Farm Implements and Machinery, CIAE Bhopal. Technical Bulletin No.: CIAE/2010/151.

Anonymous. 2012. Directory of Successful Farm Machinery in SAARC Countries. SAARC Agriculture Centre. BARC Complex, Farmgate, Dhaka – 1215 (Bangladesh).

Garg I K; Singh Surendra. 2002. Farm equipment for Punjab agriculture. Department of Farm Power & Machinery, Punjab Agricultural University, Ludhiana.

Pandey M M; Ganesan S. 2005. Farm Mechanisation Package for Dryland Agriculture, Technical Bulletin No: CIAE/FIM/2005/117. Central Institute of Agricultural Engineering, Bhopal.

Pandey M M; Majumdar K L; Singh Gyanendra; Singh Gajendra. 1997. Farm Machinery Research Digest. Technical Bulletin No. CIAE/97/69, Central Institute of Agricultural Engineering, Bhopal, 328 p.

Singh Surendra. 2007. Farm Machinery – Principles and Applications. Directorate of Information & Publication of Agriculture, Indian Council of Agricultural Research, Krishi Anusandhan Bhawan-I, Pusa Campus, New Delhi.

Singh Surendra; Pandey M M. 2008. X Plan Achievements (2002-2007). AICRP on Farm Implements and Machinery, CIAE Bhopal. Technical Bulletin No.: CIAE/2008/137.

6 Fertilizer Application Equipment

Fertilizers are required where soils are deficient in plant food elements. When land is planted to crops over a long period of time, the plant food elements slowly get reduced and productivity of crop goes down. Sandy soil looses plant food elements rapidly because heavy rainfall or application of irrigation water leaches these elements out from the soil. Adding chemical fertilizer increases fertility of soil. Principally nitrogen, phosphorous and potassium are added to the soil to promote greater yields. Uniform distribution and proper placement of fertilizer in the soil have become increasingly important factor in providing maximum crop response at minimum cost. Placing the fertilizer near the seeds at the time of sowing results in more effective utilization of nutrients. In the row crops side dressing of fertilizer is done which has immediate benefit when placed in the moist soil within the root zone. The effective use of N, P and K by crop is of utmost importance. Fertilizer use efficiency is seldom high and depends on method of application. The fertilizer use efficiency ranges from 20 to 50 percent for nitrogen, 10 to 20 percent for phosphorus and 80 to 90% for potassium (Goswami & Kamath, 1982).

In India fertilizers are mostly used in solid form. These fertilizers are either broadcasted over the surface or drilled into the soil. The use of fertilizers through irrigation water and foliar application has also been adopted recently. Broadcasting of granular fertilizer by hand leads to a non-uniform distribution and low fertilizer use efficiency. Drilling or band placement of fertilizer ensure uniformity of distribution and accuracy of placement which lead to better plant response and growth and better fertilizer use efficiency. Fertilizers are applied at different stages: partly at the time of sowing/planting or sometime even before, and partly in one-two dozes in the standing crop in the form of side or top dressing. Therefore, the choice of the method of application depends on time of application during crop growth, soil, crop, fertilizer type, quantity to be applied, availability of water and application machinery.

Fertilizer can be applied to the soil either by broadcasting or by drilling with the help of mechanical device. The uniformity of distribution is difficult to be maintained by broadcasting. Urea is highly hygroscopic and has a tendency to leach down and other fertilizers remain on the surface without being available to the crop. Thus, metering and placing of proper amount of fertilizers is important. Proper placement and uniform distribution of fertilizer can only be achieved with the use of fertilizer drills. Good fertilizer application equipment helps to apply the fertilizer at the recommended rate, at the right time and at the right place. It also avoids wastage, reduces the cost of operation and improves fertilizer use efficiency and crop yields. The uniformity and ease of drill ability of fertilizers through openings depends on flow ability, clod formation under the influence of agitators, moisture, size and shape of the particles and bulk density. Any slight change in air humidity changes shape, makes them clody, opts to bridging and thus, changes the flow behaviour through openings. To prevent bridging for uniform and free flow of fertilizers, agitator of different shapes (Fig. 6.1) is located inside the fertilizer box. Different types of agitators viz. star, rod type, circular rubber or canvas flaps, vane and notch discs are commonly used. Circular shape openings give higher discharge rate than square openings. In general star type agitator give higher discharge rate than other types of agitators.

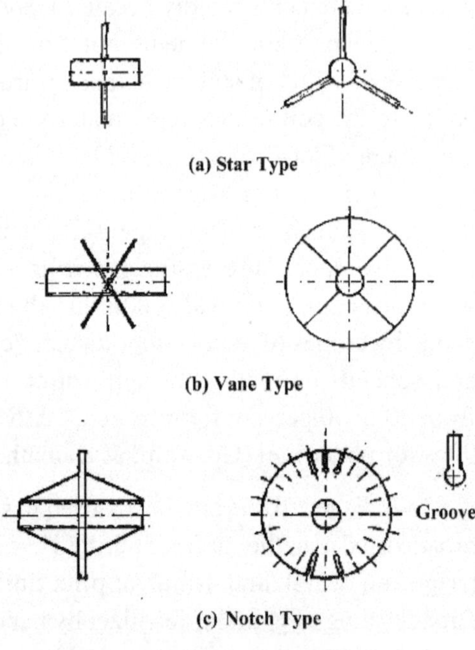

Fig. 6.1: Different types of agitators used in fertilizer drills.

Selection of fertilizer application equipment

The following points need to be kept in mind while selecting fertilizer application equipment:

- suits the source of power available,
- easy to adjust, operate and maintain,
- have a positive and accurate metering device,
- have proper type of furrow openers,
- have a proper calibration arrangement and index plate to adjust the application rate,
- hopper, metering device and delivery tube is made of non-corrosive ma terial,
- have an instant closing orifice, and
- easy to clean left over fertilizer

Dry fertilizer application equipment

Dry fertilizer is applied practically to most of the crop in all types of field conditions. The rate of application varies according to the type of fertilizer, crop, soil and irrigation levels etc. Some of the application methods for dry fertilizer are (i) broadcast the fertilizer before ploughing or place it at ploughing depth, (ii) broadcast the fertilizer and mix into the soil before sowing, (iii) deep placements of fertilizer with deep plough viz. chisel plough, sub soiler, (iv) apply fertilizer during drilling or planting, and (v) apply fertilizer through sides (side dressing) in row crops. In general, distributors used for applying dry fertilizers may be classified as those that broadcast the material onto the surface of the ground and those designed for placing the fertilizer in rows or bands below the surface. Drop type of broadcaster is suitable for spreading either fertilizer or lime. Furrow opener is also provided in some of such equipment for band placement below the soil surface and can be used for side dressing of row crops. These may be either pull type or mounted type implements. These are also available as an attachment to various implements. Centrifugal type of fertilizer broadcaster uses hopper that meters fertilizer and distributes it laterally over the field with the help of one or two horizontally rotating ribbed discs. This type of broadcaster can also be used for broadcasting small seeds in the field. Most of these units employ centrifugal spreading primarily because of the compactness and simplicity in comparison to throw type distributors. In this type of broadcasters, non-uniform lateral distribution is a problem, which has been solved by taking proper left and right turnings to overlap the distribution pattern and make it as uniform as possible. These fertilizer broadcasters may be manually operated or tractor operated.

Manually operated fertilizer broadcaster

Majority of farmers still practices the hand broadcasting of fertilizer. The principal disadvantage of hand broadcasting is the non-uniformity of distribution causing uneven growth of crop, which ultimately results in poor yield. The problem of non-uniform distribution of fertilizer in the field can be easily overcome, if manually operated fertilizer broadcaster is used (Garg and Singh, 2002; Anonymous, 2012). It can be used for broadcasting granulated fertilizer viz. Urea, DAP etc. The effective width of spread is about 6.4 m. The manually operated fertilizer broadcaster consists of a cylindrical hopper box made of either sheet metal or special grade plastic with tapered bottom, fertilizer-metering mechanism, spreading disc, transmission unit, agitator, handle and strap (Fig. 6.2). All these components are fitted over a lightweight rigid frame. Generally aluminum alloy is used for this purpose. The hopper bottom normally has a circular hole for metering fertilizer. This hole is having a sliding gate, which reduces or increases the notch opening area with the help of a lever. The ratio between the diameter and height of the hopper is kept in the range of 0.8 to 1.25. The hopper can be covered with a lid with a peeping hole of at least 75 mm in diameter; to see the quantity of fertilizer in the hopper during operation. The hopper should be sufficiently strong and should not buckle when filled fully with fertilizer. The capacity of hopper varies between 12 to 15 liters.

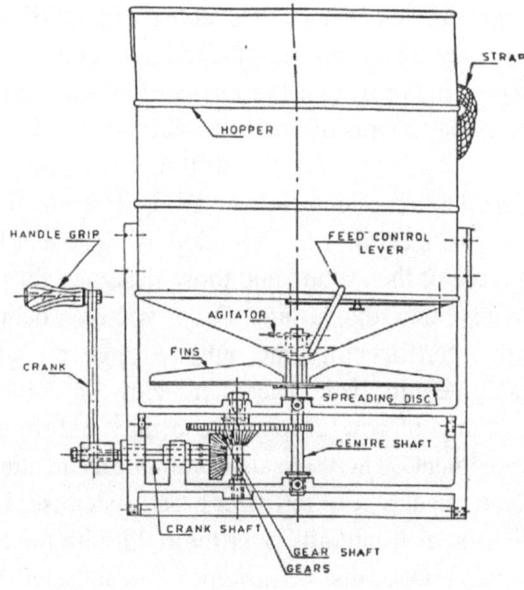

Fig. 6.2: Manually operated fertilizer broadcaster

Spreading disc is made of aluminum alloy or MS sheet. It is mounted at the bottom of the hopper, which may have 6 to 8 equally spaced blades/fins. The spreading disc of 27 cm diameter is provided with a vertical clearance of at least 30 mm from the hopper bottom. This is rotated with the help of transmission unit and it spreads the fertilizer falling over it from notch opening of hopper uniformly in the field. Feed control mechanism with loading device is provided to control the flow of fertilizer through the aperture. This mechanism is controlled by a lever provided on the side of the hopper and can be moved easily to the left or right from mean position to have required notch-opening area. An index pointer is also provided on the hopper, which correspond to the notch opening position on the chart having marking ¼, ½, ¾ and full. Agitator is provided to avoid the clogging of the aperture and for feeding the fertilizer to the aperture properly near the orifice in the hopper. This agitator is provided just above the aperture and a suitable vertical clearance is provided between the agitator and the aperture.

Transmission unit, which rotates the spreading disc, consists of set of gears and shafts with bearing and bushes. The transmission unit rotates the spreading disc at a peripheral speed of 450-550 cm/s. The set of gears are so arranged that it converts the rotation of handle into right angled rotation of spreading disc. Handle rotates the pinion which rotates a helical gear over which spreading disc is mounted. The speed ratio is kept in such a way that the disc rotates at much faster rate than the handle rotation. Normally, two straps of suitable length are provided in the fertilizer broadcaster to help easy carriage of broadcaster. The length of strap can be adjusted as per requirement of the operator. Hence, operator can have the broadcaster mounted at his front at a convenient height. Some of these straps are also provided with cushioning pads at shoulder position. This cushion is covered with cotton or leather. This cushioning in the strap reduces the fatigue of the operator. The total weight of the machine is kept as low as possible, so that it can be easily carried in the field by the operator. Normally, the weight of empty machine is around 5 kg. Fertilizer is filled in the hopper and broadcaster is mounted at the front of the operator. By rotating the handle fertilizer from the notch-opening fall over to the rotating disc, which spreads the fertilizer evenly into the field (Fig. 6.3). The operator is asked to move in the field at its own normal speed. The spread pattern of this type of broadcaster is generally skewed towards right of the line of travel. The skewness is a function of drop ratio. The skewness can be reduced to a minimum with the help of proper notch opening location, height of drop and speed. Else, taking proper overlapping can solve the problem of skewness.

Machine performance for uniform distribution of granular fertilizer is very good. Broadcaster in one operation can apply the recommended dose of urea uniformly. Machine has a capacity to cover about 0.8 ha/h whereas area covered manually is only 0.36 ha/h.

Courtesy: M/s. National Agro Industries Ludhiana
Fig. 6.3: Manually operated fertilizer broadcaster in use in field

Tractor operated fertilizer broadcaster

This type of fertilizer broadcaster also works on the similar principal, as in the manually operated broadcaster (Singh, 2007). The required rotations to the spinning disc are provided through tractor PTO shaft (Fig. 6.4). It consists of an MS sheet hopper of conical shape. At the bottom of the conical hopper there is an adjustable gate, through which fertilizer is allowed to pass. Amount of fertilizer dropped over to the spinning disc can be varied to obtain desired fertilizer rate. The disc is rotated with the help of shaft mounted at the center of the disc. The other end of disc shaft is connected to the gearbox shaft. On the other side of the gearbox, there is a splined shaft that can be connected to tractor PTO. Gearbox is provided to reduce the speed and increase the torque of disc. The required rpm of disc is achieved by providing suitable reduction gear. Fertilizer broadcaster is mounted at the rear of the tractor. As the tractor PTO moves the disc rotates and the fertilizer from the gate falls over to the disc and spread uniformly towards both left and right side of the tractor. The pattern of fertilizer distribution is not as skewed as in the case of manual broadcaster. A tractor mounted fertilizer broadcaster with or without oscillating spout is also used for top dressing of granular fertilizers (Fig. 6.5). It has a hopper capacity of 2.5 to 3.0 quintals of granular fertilizer. It spreads the fertilizer over a swath width of 6 to 8 m and requires 35 hp tractor. It maintains good uniformity of distribution and is used under different soil conditions with adequate moisture and where irrigation facilities are available. It can also be used for top dressing application.

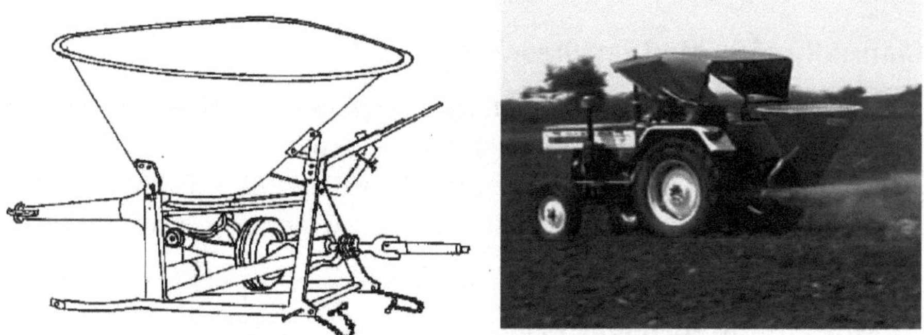

(*Courtesy*: Farm Implements (India) Pvt. Ltd., Chennai)
Fig. 6.4: Tractor mounted fertilizer broadcaster.

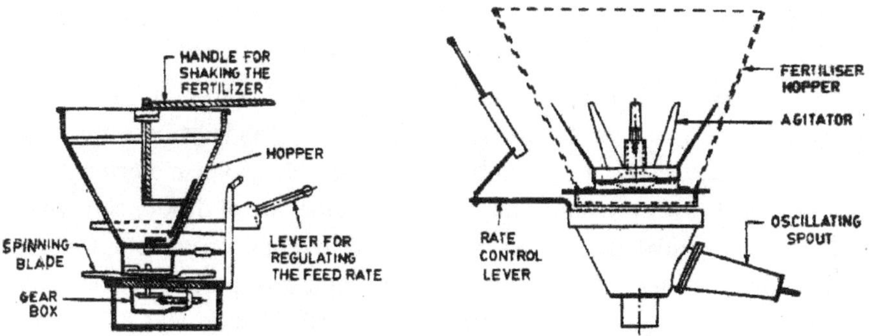

Fertilizer spreader without oscillating spout   Fertilizer spreader with oscillating spout

Fig. 6.5: Tractor operated fertilizer broadcaster with and without oscillating spout.

Fertilizer drills

Most of the seed drills and planters have arrangement for placement of fertilizer in soil at desired depth at the time of sowing/planting (Singh, 2007; Pandey *et al.*, 1997). However, for additional dosage of fertilizer at different stages of a crop, fertilizer is required to be applied separately. Depending on the needs of the plants for particular nutrients, the fertilizer is broadcasted over the field, placed in the soil at a depth of seed or deeper, or distributed at one or both sides of the rows of crop as close to the plants as may be required for nutrition. For fertilizer applicators, uniformity of application and maintenance of fertilizer rate are the two most important performance parameters. The machine should have wide range of controls for feed rate and adjustment to take care of varying physical properties of fertilizers.

Manually operated fertilizer drill

Manually operated seed-cum-fertilizer drills for small holdings have gravity type metering device with an adjustable orifice to control the fertilizer rate. An agitator is also provided to ensure uniform flow of fertilizer. Two persons are required to operate the drill; one pulls the drill and other guides it in the field. It can cover 0.3 to 0.4 ha/day.

Animal-drawn fertilizer drills

The simplest device for placement of seeds and fertilizer is an attachment to the country plough called 'Pora' method of sowing. It consists of a funnel with tube attached to a country plough. One man guides the plough and other drops the seed/fertilizer in the funnel. In some animal-drawn 'desi' plough provision for agitator in fertilizer box is made. The drill uses gravity or fluted feed metering devices to control fertilizer rate. It can cover 0.4 to 0.6 ha/day. Several types of 2 to 5 rows animal drawn drills with variation in metering devices and furrow openers are also available. The seed and fertilizer are placed in different boxes. Mostly fertilizer box is having gravity type metering device with an adjustable opening and an agitator. Agitators are of disk or spur wheel type. An agitator prevents fertilizer from bridging over the opening and ensures uniform flow. The fertilizer is conducted through transparent plastic tubes, which helps in detecting the clogging of fertilizer in tube and also avoids corrosion. Its field capacity varies from 1 to 1.5 ha/day.

Tractor operated drills

Several types of tractor operated drills and planters with fertilizer attachment are available in India. It uses vertical disc type metering device with grooves on the periphery for both seeds and fertilizers. These machines are comparatively cheaper, more durable and easy to operate. It covers 2.5 to 3.0 ha/day at the forward speed of 4-5 km/h. A seed-cum-fertilizer drill using external fluted feed roller for metering fertilizer has also become popular. The drill has an average field capacity of 2.5-3.0 ha/day. The drills usually places fertilizer in band on one side of the seed at the same or slightly higher depth than the seed. Agronomic research have shown that the fertilizer placement in a band about 2.5 cm to the side and 2.5 cm deeper than seed lead to about 8 to 10% increase in yield of wheat and other cereal crops.

Fertilizer drill for dryland farming

The simplest device for line sowing of cereals and band placement of fertilizer used in dry farming areas, is an attachment to country plow having a bamboo or metal pipe with a bowl (Pandey and Ganesan, 2005). Such type of drills is available in one to three furrows. Seeds and fertilizer are metered and dropped by hand. The accuracy of metering depends on the operator's skill and experience. However, this method of placing seed and fertilizer is certainly superior to hand broadcasting and leads to better fertilizer use efficiency. For the light and medium soils, single-row animal drawn and two-row tractor mounted ridger type seeder with fertilizer attachment are also available. It has provision to sow the seeds in the furrow for the rabi crops when the moisture is scarce and on the ridges during kharif. The furrows during kharif season help to harvest the rainwater and promote better plant growth. It has been found that bed planting in low rainfall area help in establishing good crop stand and increases the crop yield significantly. This equipment has provision to drill the fertilizer in a band on one side of the seed.

Common type of furrow openers

The seed-cum-fertilizer drill used in India uses different type of furrow openers depending upon the type of soil, surface condition and irrigation water availability (Fig. 6.6), Singh (2007). The type of furrow openers used for normal irrigated soils, clod-forming soils and rainfed soil differs (Table 6.1). The split-boot furrow opener, commonly known as single-hoe double pore opener, is the most common opener used on animal as well as tractor-drawn drills in northern India because of its simple design, low cost and low draft. This type of furrow opener places the fertilizer in a band about 2.5 cm to the side and about 2.5 cm deeper than seed.

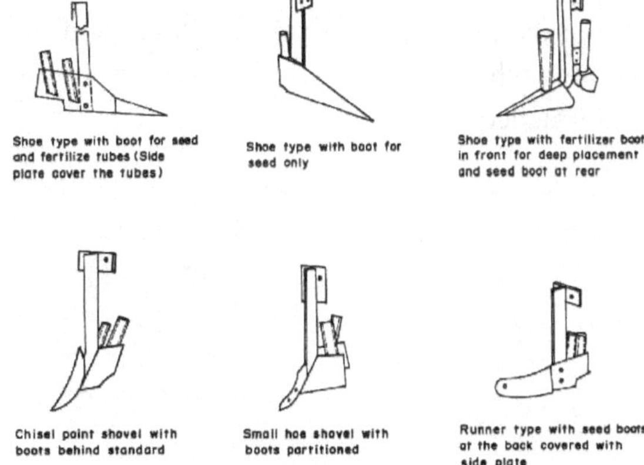

Fig. 6.6: Common types of furrow openers used in fertilizer drills

Table 6.1: Type of furrow openers used in fertilizer drills

Type of Furrow Opener	Description	Suitability
Hoe type	Single or double pointed shovel with one or two pores/boots.	Suitable for light, medium soils. Soil free of excessive trash has good penetration.
Stub or full covered runner	Resembles a curved sword with a thin sharp cutting edge with single or double boots at the rear.	Widely used on row crop planters. Suitable for shallow sowing. Sharp blade cuts through the clods and sod. Low draft and minimum soil disturbance.
Single disk opener	One disk slightly curved, fastened to the boot and set to run at an angle.	Good penetration, cuts the trash and does not clog.
Double disk opener	Two disks facing each other placed at a slight angle.	Suitable for deep sowing at relatively higher speeds. Ideal for sowing small seeds in trashy seedbeds.
Chisel-type furrow opener	A body with bar shape has the shape point projecting over the shoe.	Especially suited to very hard and cloddy soils typical in the black soil belt in the rainfed areas.

Common type of ground wheels

Ground wheels used on seed-cum-fertilizer drills support the drills for proper movement in the field as well as power the metering devices (Singh, 2007). Power is transmitted from ground wheel to the metering devices through chain and sprocket. The type and size of ground wheel depend on the soil type, surface condition and the type of drill/planter (Fig. 6.7). Plain wheel is used to support the drill, whereas wheel with lugs is used to power the metering device in soils having clod and trash and stones and also in sticky soils. Pegs on wheel reduces wheel slippage in cloddy, trashy and sticky soils and improves performance.

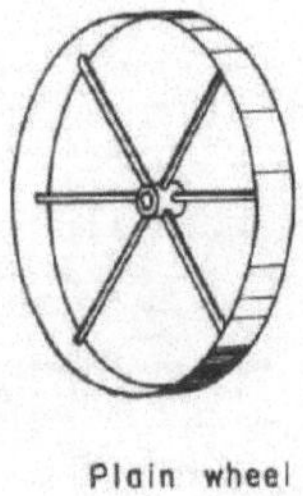

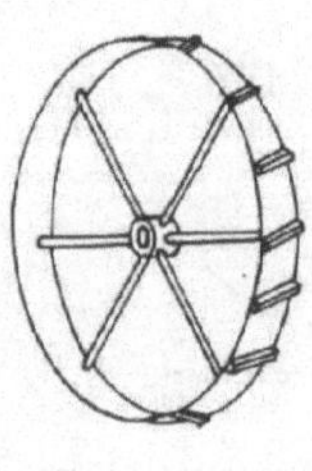

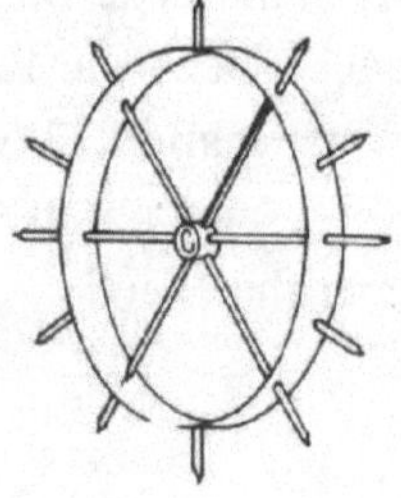

Fig. 6.7: Common type of ground wheels

Metering devices

Fertilizer drill uses mainly two types of metering devices namely: gravity feed type with agitators and positive feed type (Singh, 2007; Singh and Verma, 2009; Pandey *et al.*, 1997). The variation in fertilizer placement in rows and between rows depends on the type of metering devices used in the machine. The positive metering devices with vertical wheel having grooves on the periphery and helical fluted rollers have low intra row variations in the rate of fertilizer dropped as compared to the gravity feed metering device with adjustable orifice and agitator. The metering device with vertical wheel having grooves on the periphery has the minimum intra row variation of 13-16 percent as against 63-66 percent in the gravity feed device. This variation is lowest (6-8%) in case of helical fluted roller type metering device. There are different types of metering devices used to meter the granular fertilizer to get uniform metering action under varying conditions (Fig. 6.8). These common metering devices are:

Star wheel: It is an old metering device and has been in use for metering fertilizer from years together. It is used in seed drills and also in some planters. There are star wheels, which carrying fertilizer through gate opening from hopper into delivery compartment. The star wheel is rotated in horizontal plane. The fertilizer trapped between the teeth of wheel falls into the delivery tube by gravity. The material carried on the top of the wheel is scrapped off and it does not come into delivery tube. Raising or lowering a gate just above the star wheel can control the rate of discharge or fertilizer application rate. Speed ratios can be altered through transmission unit. The star wheels are driven through transmission unit which consist of set of gears from feed shaft below the hopper and gets the drive through ground wheel.

Rotating bottom plates: In this type of fertilizer metering device there is a horizontal rotating bottom plate which is fitted over to a stationary bottom ring of hopper base. This type of attachment is used in row crop planter. An adjustable gate fitted over side outlet can control fertilizer rate.

Auger feed metering device: It uses an auger to meter the required amount of fertilizer. Auger is closely fitted into tube. A star wheel type agitator is also fitted in the hopper to avoid clogging of fertilizer in the auger. From the hopper auger carries the fertilizer into tube and drops in the delivery tube.

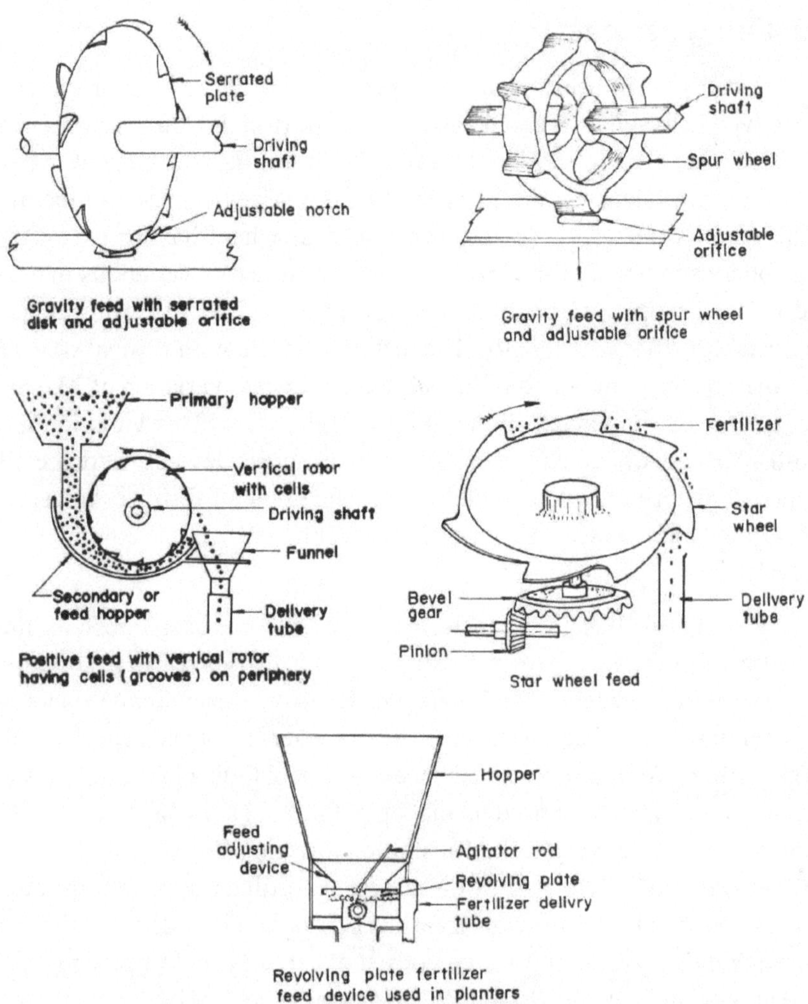

Fig. 6.8: Common types of fertilizer metering devices.

Edge cell feed type: It consists of a vertical rotor fitted in the hopper, having cells on its entire periphery. Number of such metering devices is installed over to entire length of hopper to meter granular fertilizer. Varying the rotor speed can vary the discharge rate.

Stationary opening metering devices: This type of metering device uses a stationary opening at the bottom of the hopper. In this opening a moving gate is provided to vary the size of opening to control the rate of fertilizer application. Sometimes an agitator is also provided to avoid any clogging of opening with fertilizer. In this type of fertilizer broadcaster, the bottom is made conical for easy flow of material.

Fertilizer application equipment for wetland

The traditional practice of surface broadcasting in wet land field is wasteful as 20-40 percent of the broadcast fertilizer is lost primarily by ammonia volatilization from the surface water. Deep placement of fertilizer results in 10-20% increase in grain yield over conventional broadcast practice. Suitable fertilizer application equipment for deep placement has been developed at IRRI that could deep place fertilizer in wet paddy fields. Push type fluted roller applicator has been developed for placement of prilled urea. Prilled fertilizer is metered by fluted roller and gravity dropped into the opened furrow. The fertilizer is covered as the applicator is pushed forward in the field. Push type gravity feed liquid applicator and pull type band placement liquid applicator have been developed for placement of dissolved urea (Khan, 1982). A manually operated push type super granule applicator (furrow opener type) is also used for placing of fertilizer (Fig. 6.9). It consists of a hand actuated plunger type metering device, fertilizer hopper and plunger tube. The plunger tube receives a continuous supply of super granules from the hopper through a connecting feed tube in which a concentric tubular agitator oscillates up and down. The machine could deep place and cover fertilizer adequately and could fertilize one hectare in 8 days.

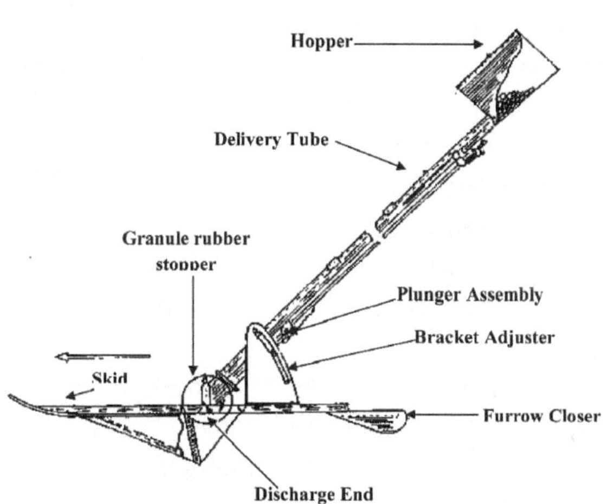

Fig. 6.9: Push-type super granule applicator

Manure application equipment

In order to increase the fertility of soil, manure is applied. This manure may be barnyard manure or compost manure. The equipment used for transferring such manure from store to the field and spreading the same in field is called manure spreader. There are two types of manure spreaders in use. These are tractor driven spreader and ground-wheel driven spreader (Singh, 2007).

Tractor driven spreader

This type of spreader is powered with tractor PTO shaft (Fig. 6.10). Spreader consists of upper beater, lower beater, spreader and transmission unit (Singh, 2007). The upper beater is placed at the top of the machine. It beats the large flakes to break it into smaller size. Lower beater is positioned at the lowest part of the machine. It is provided to beat the flakes and break them apart. It runs in opposite direction of main wheel and it handles the hard sticky manure flakes easily. The broken manure with the help of beaters is spread over to field. Spreading of manure is done through spreader consisting of spiral auger like blades, which designed in such a way that; half of the blade turns the manure to the right and other half to left. The PTO of tractor is connected to the input shaft of gearbox fitted at the rear end. Gearbox is having arrangement by which lower beater; upper beater, feed conveyor and drive spreader are connected. It contains a feed conveyor, which conveys the manure from the front part of the container to the rear end for onward spreading in the field. The drive sprockets usually runs at very slow speed 1-5 rpm and this speed can vary with the help of selection of gear through gearbox. The feed rate of manure can be controlled with the help of stop pawl.

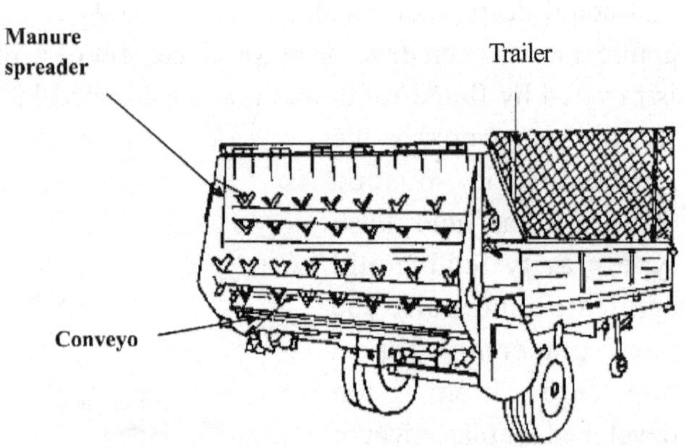

Fig. 6.10: Tractor-operated manure spreader

Tractor operated farmyard manure spreader

A two wheel tractor operated farmyard manure spreader consists of spreader and concave unit, feeding auger, slanting sheet platforms and sliding sheet to control the manure delivery rate (Fig. 6.11), Anonymous (2008, 2010). Six splines standard PTO shaft and a gear box with speed reduction ratio 2:1 are used for transmission of power to spreader and feeding auger. Shafts, chains and sprockets are used to transmit 540 rpm at spreader unit and 70 rpm at feeding auger. The trailer has length and width as 4500 ×1800 mm, wheel base 1400 mm and ground clearance of 800 mm. The manure-spreading unit is mounted below the trailer in front of pneumatic wheels. Unit consists a rotating drum and concave. The manure-feeding auger is made of mild steel spiral discs of 2 mm thickness. The diameter of disc is 250 mm. Two number slanting sheet made of G I sheet of 2 mm thickness are provided on the M S angle frame. These sheets make a "V" shape body and are used for sliding the manure from the box to the feeding auger. The slanting platforms are hinged in the bottom of trailer. A sheet sliding mechanism has been provided on the bottom of platform to control the manure delivery rate for the spreader unit. By increasing/decreasing the opening width the manure delivery rate is controlled. The manure spreader has been calibrated for determination of uniformity of spreading and application rate. A tractor operated farmyard manure spreader of two tonne capacity gives manure application rate of 4.5 – 36.3 t/ha for the manure delivery rate of 1.05 - 3.06 kg/s at the forward speed of 0.41 - 1.12 m/s. Coefficient of uniformity of manure distribution varies from 86 - 93%. The field capacity of machine varies from 0.23 to 0.58 ha/h at the forward speed of 0.41 to 1.12 m/s, respectively.

Fig. 6.11: A view of tractor operated farmyard manure spreader

Slurry distributor

Slurry consists of undiluted dung and urine with some feeding-stuffs such as 'bhusa', hay or silage. They cannot be easily handled effectively without some dilution. One of the potential advantages of handling liquid manure as slurry is that it conserves more of the fertilizer constituents, especially the nitrogen. The slurry can further be diluted normally into 3:1 ratio, which can be handled effectively by a tank spreader (Singh, 2007). The machine consists of either vacuum-filled or pump-filled mechanism. Even the diluted slurry needs continuous mixing through a shaft-operated centrifugal pump, which is installed in a sump. The pump may incorporate a chopping mechanism to make fibrous material more pumpable. The vacuum-fill pressure discharge type slurry tankers work satisfactorily, if the slurry is free flowing liquid (Fig. 6.12). The machine includes a slurry tank, vacuum pump, regulator, inlet hose and spreading arrangement. The gravity-type mechanical discharge type slurry tanker can handle thick slurries provided it can be filled in the tanker. This machine is dual-purpose type: has spreading rotor for solid slurry and an impeller for liquid slurry. They are mounted on separate drive and can be alternatively used according to need. Because of atmospheric pollution considerations and the machine throws small particles for long distances, its use is not recommended near residential area.

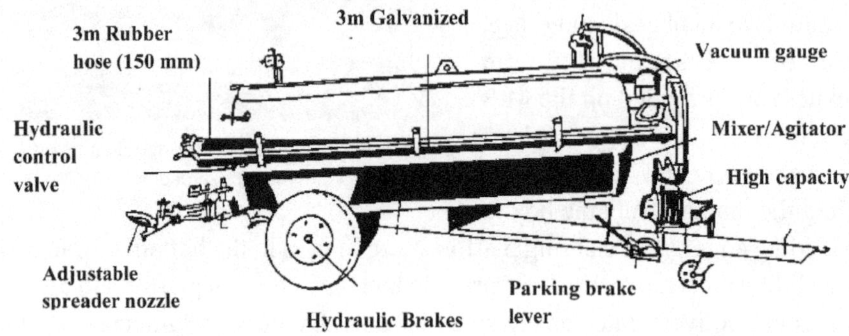

Fig. 6.12: Vacuum-fill, pressure-empty slurry distributor

TNAU Power tiller operated manure spreader

The TNAU, Coimbatore has developed a power tiller operated manure spreader (Fig. 6.13) to expedite the operation and reduce human drudgery (Anonymous, 2006). Use of this attachment increases the annual use of the power tiller and facilitates timely completion of manure spreading. The unit consisted of main frame, manure tub, conveying mechanism, spreading mechanism and adjustable rear aperture. A front mounted adjustable parking stand is provided to support manure spreader for convenient hitching with the power tiller. The chain conveyor with feeder plate is mounted over the upper part of the manure tube. The farm yard manure is conveyed from front to rear portion of the manure tube by chain conveyor. Drive to the chain is provided by a ground wheel and the spreader assembly is mounted at the

Fig. 6.13: TNAU Power tiller operated manure spreader

rear end. The main rotary spreading disc of 700 mm diameter has 8 nos. of straight vanes riveted on it. It is driven by a hydraulic motor, and operated at 200 rpm by adjusting the hydraulic control valve. Quantity of manure spread is controlled by an adjustable flap at the rear of the manure tube. The field capacity of the manure spreader is 0.9 ha/h. The savings in cost and time of operation are 85% and 94% respectively as compared to manual method of manure spreading. Width of coverage is 6 m. The manure application rate is 16.56 t/ha.

Fertilizer applicator and earthing up equipment for sugarcane

Weeding and interculture is very important in sugarcane cultivation. Mechanical devices have been found very effective for control of weeds and creating favourable environment for sugarcane crop. With the use of different types of interculture, weeding equipment advantages of secondary tillage are achieved. In sugarcane in addition to basal doze of fertilizer application top dressing is also done for which tractor operated fertilizer applicators-cum-earthing up equipment have been developed (Fig. 6.14), Singh (2014). It consists of two rotary hubs with blades, two fertilizer boxes, frame, two earthing up devices and power transmission system. Row spacing can be adjusted. It can be operated in the cane rows planted at 90-120 cm spacing for interculture purpose.

Fig. 6.14: Tractor operated fertilizer applicator cum earthing up equipment

References

Anonymous. 2006. Research Highlight. AICRP on Farm Implements and Machinery, CIAE Bhopal. Technical Bulletin No.: CIAE/FIM/2006.

Anonymous. 2008. Research Highlight. AICRP on Farm Implements and Machinery, CIAE Bhopal. Technical Bulletin No.: CIAE/2008/141.

Anonymous. 2010. Research Highlight. AICRP on Farm Implements and Machinery, CIAE Bhopal. Technical Bulletin No.: CIAE/2010/151.

Anonymous. 2012. Directory of Successful Farm Machinery in SAARC Countries. SAARC Agriculture Centre. BARC Complex, Farmgate, Dhaka – 1215 (Bangladesh).

Garg I K; Singh Surendra. 2002. Farm equipment for Punjab agriculture. Department of Farm Power & Machinery, Punjab Agricultural University, Ludhiana.

Goswami N M; Kamath M B. 1982. Fertilizer use in multiple cropping systems and measures to increase efficiency. Proceedings of FAI-Northern region Seminar on Accelerating the Pace of Fertilizer Consumption, New Delhi. Pages 17-34.

Khan A U. 1982-84. Deep Placement Fertilizer Applicator Project, Semi-annual Progress Reports No. 1-5. Agricultural Engineering department, IRRI, Manila, Phillippines.

Pandey, M.M. and Ganesan S. 2005. Farm Mechanisation Package for Dryland Agriculture, Technical Bulletin No: CIAE/FIM/2005/117. Central Institute of Agricultural Engineering, Bhopal.

Pandey, M.M., Majumdar, K.L., Singh Gyanendra; Singh Gajendra. 1997. Farm Machinery Research Digest. Technical Bulletin No. CIAE/97/69, Central Institute of Agricultural Engineering, Bhopal, 328 p.

Singh, P.R. 2014. Souvenir, Tractor and Agricultural Machinery Manufacturers' Meet (TAMM 2014). Held at IISR Lucknow during February 8-9.

Singh, Surendra. 2007. Farm Machinery – Principles and Applications. Directorate of Information & Publication of Agriculture, Indian Council of Agricultural Research, Krishi Anusandhan Bhawan-I, Pusa Campus, New Delhi.

Singh Surendra; Verma S R. 2009. Farm Machinery Maintenance & Management. Directorate of Information & Publication of Agriculture, Indian Council of Agricultural Research, Krishi Anusandhan Bhawan-I, Pusa Campus, New Delhi.

7 Plant Protection Equipment

The application of agricultural chemicals for control of insects, pests, fungi, weeds and plant diseases plays an important role in crop production. Before 1980 few chemicals were known for control of plant pests. The chemical pesticides have played an important role in the rapid advancement of agricultural production. Chemical pesticides are getting popular due to non availability of labours especially for weeding operation during kharif season. The chemical pesticides and herbicides are applied with the help of different types of sprayers and dusters. The spraying equipment may be used for (i) application of herbicides in order to reduce competition from weeds; (ii) application of protective fungicides to minimize the effects of fungal diseases; (iii) application of insecticides to control various kinds of insect's pests; and (iv) application of micronutrients such as manganese or boron (Singh and Kandoria, 1999; Pandey *et al.,* 1997; Singh, 2007). Spray application involves different sizes of the droplet application according to which spray application volume changes. According to droplet size the spray application may be classified as Aerosols (< 50 micron); Mist (51 - 100 micron); Fine spray (101 - 200 micron); Medium spray (201 - 400 micron), and Coarse spray (> 400 micron).

The main function of a sprayer is to break the liquid into droplets of effective size and distribute them uniformly over the surface or space to be protected. Another function is to regulate the amount of insecticide to avoid excessive application that might prove harmful or wasteful. A sprayer that delivers droplets large enough to wet the surface readily should be used for proper application. Extremely fine droplets of less than 100 micron size tend to be diverted by air currents and get wasted. Some estimates have suggested that up to 80% of the total pesticide applied to the plants may eventually reach the soil. Different designs of spraying equipment have been developed for different types of applications and field and crop conditions. These sprayers are classified as i) hand operated hydraulic sprayers, ii) foot operated hydraulic sprayers, iii) power operated hydraulic sprayers, iv) air carrier sprayers, v) spinning disc sprayers, vi) dust and grannule application equipment, and vii) aerial application equipment.

Spraying techniques are classified as high volume (HV), low volume (LV) and ultra low volume (ULV), according to the total volume of liquid applied per unit of ground area. Initially high volume spraying technique was used for pesticide application but with the advent of new pesticides the trend is to use least amount of carrier or diluent's liquid. In spraying, the optimum droplet size differs for different types of application. Fine droplets are required to control insects, pests or diseases and bigger size droplets for application of herbicides, etc. The greater the number of fine droplets produced by the device better will be deposition on target area. The size of droplet is important as it affects drift and penetration distance of droplets towards the target. Hence a compromise is to be made to prevent drift, achieve wide coverage of plant or target area and more penetration.

Types of spraying equipment

Different designs of spraying equipment have been developed for different types of applications and field and crop conditions. Manually operated hydraulic sprayers viz. knapsack sprayers, twin knapsack sprayers, foot sprayers, hand compression sprayers; air carrier sprayers such as motorized knapsack mist blower cum duster (LV) and centrifugal rotary disc type sprayers are specially suitable for spray applications in crops (Singh and Kandoria, 1999; Pandey *et al.*, 1997; Singh, 2007). The present trend is to apply concentrated pesticides by means of low and ultra low volume sprayers. This has been possible through the development of better formulations and special nozzles. New Controlled Droplet Application (CDA) atomizers require less than 15 l/ha of spray mixture and are easy to operate. Although new type of spray atomizers is available but the correct chemical formulations are commonly not produced. Hence their use is limited to particular crop, pest or disease. The instructions of the manufacturer should be carefully read whether a formulation is recommended for ultra low or low volume application.

Hand sprayer

Fig. 7.1: Hand sprayer

The hand sprayer is a small capacity pneumatic sprayer (Fig. 7.1). It consists of chromium plated brass tank having a capacity of 0.5 to 3 litres (one litre is more common) which is pressurized by a plunger pump. The air pump remains inside the tank. The sprayer has a short delivery tube to which a cone nozzle is attached. In some models, the nozzle is attached at the top of the tank with flow spring actuated lever, which regulates the flow of the spray liquid. For spraying, the tank is usually filled to three-fourths capacity and pressurized by air pump. The compressed air causes the agitation of the spray liquid and forces it out, on operation of the trigger or shut off type valve. Usually the chemicals with suspension characteristics cannot be effectively sprayed with this type of sprayer. For spraying wettable powders the sprayer is shaken frequently to prevent settling of the chemical. For operation, the spray nozzle is directed to the target after charging. It is fitted with mist spray nozzle with gooseneck bend. The pump assembly is made of brass and operated by one person. It is ideal for small nurseries, rose plants, kitchen gardens and spraying wettable insecticides and fungicides.

Stirrup pump sprayer

The stirrup pump sprayer commonly used for Mosquito control, is quite popular among the small farmers and vegetable growers due its simplicity, low cost and ease of operation. It is also called bucket sprayer, as the pump always remains submerged in a bucket containing the spray liquid (Fig. 7.2). It consists of a double action pump, seamless brass tube barrel, adjustable stirrup foot rest, angular delivery spout,

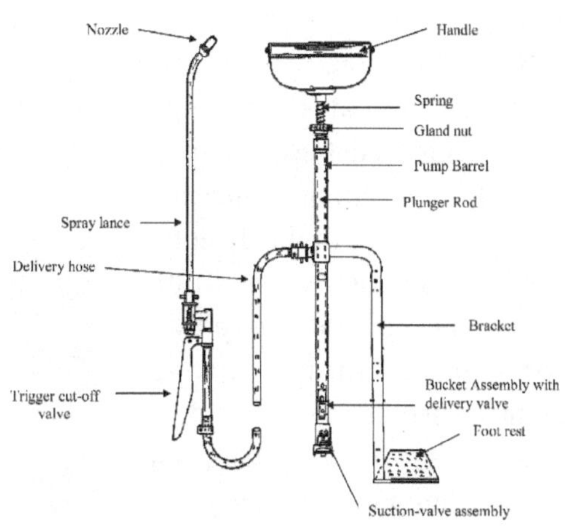

Fig. 7.2: Single-barrel stirrup pump sprayer

relief valve in the foot valve, plunger, shaft fitted with travel limiter, D type handle, brass balls, detachable gooseneck bend in the spray lance, spray lance fitted with nozzle, delivery hose and trigger cut-off valve with strainer. For operation, the barrel fitted with pump is placed in a bucket containing spray liquid and one person operate the pump by placing his foot on the foot stirrup and moves the pump shaft fitted with D type handle. The other person holds the spray lance and directs it towards the target. All the major parts are made of brass. Usually a flat fan spray nozzle is used with the sprayer. The sprayer is provided with 5 m delivery pipe. The field capacity of the sprayer is 0.3 ha/day. It is used for spraying in orchards, nurseries, flower crops, vegetable gardens etc.

Hand compression sprayer

Hand compression sprayers are either pressure retaining or non-pressure retaining type. The pressure retaining type has an advantage that air charged once may last for weeks, but requires sturdy tank and high pressure, therefore these are not in common use. Non-pressure retaining type is the most commonly used hand compression sprayer. Like other sprayers, it consists of an airtight metallic tank, air pump; lance fitted with trigger type or shut off valve, gooseneck bend, a pair of shoulder-mounted straps and nozzle (Fig. 7.3). All the parts are made from brass alloy and the tank is fabricated to withstand high-pressure up to the order of 18 kg/cm^2. Tank capacity ranges from 9 to 14 litres depending upon model. It is a low volume sprayer. For operation, the tank is filled to three fourths of its capacity and pressurized by hand plunger pump, which remain inside the tank or from a compressor. The pressure inside the tank is usually maintained at 3-4 kg/cm^2. The operator mounts the sprayer on his back securing it by shoulder straps and operates the trigger valve, which enables the spray liquid to flow through lance and nozzle. The lance is directed towards the target. A single person can operate the sprayer. For maintaining proper atomization of the spray liquid, the tank requires frequent pressurization. The discharge and atomization decreases with

Fig. 7.3: Hand compression sprayer

decrease in pressure. The hand compression sprayer is used in kitchen gardens, nurseries, vegetable gardens, flower crops and field crops.

Rocker sprayer

The rocker sprayer is a long lever high-pressure sprayer designed for operation with one or two lances. The complete assembly is mounted on a wooden board, which is held to the ground by the foot of the operator (Fig. 7.4). The sprayer consists of a single or double acting piston pump for developing high pressure, an air chamber, spray lance with shut off valve and strainer, 5 m suction line fitted with strainer and delivery line. The principal components are made from brass alloy. The lance is fitted with gooseneck bend and nozzle and the length of lance may vary from 60 to 90 cm. The pump is operated with long lever to and fro in a rocking motion which sucks the liquid from the inlet pipe submerged in the spray liquid. The other person holds the lance and directs the spray chemical to the target. If two lances are used, then it may require in all three persons for the spraying operation. With high jet spray gun or bamboo lance the spray chemical can be delivered to a height of up to 10 m. It is used for spraying on tall trees like coconut, areca nut, sugarcane, rubber plantations, orchards, vineyards and field crops, vegetable gardens, flower crops etc.

Fig. 7.4: Rocker sprayer

Foot sprayer

The foot sprayer is one of the ideal and versatile sprayers used for multipurpose spraying jobs. The principle of working is similar to the rocker sprayer. The sprayer consists of a pump operated by the foot lever, suction hose with strainer, delivery hose, spray lance fitted with shut off pistol valve, gooseneck bend and adjustable nozzles (Fig. 7.5). The pump barrel is mounted on a steel frame, which provides it stability when placed on the ground. It has a provision of two strong springs, which retract the foot lever to its original position after each pumping stroke. The sprayer does not have inbuilt tank, therefore an additional storage device or container is required to store the spray liquid in which the strainer of suction hose remain submerged. It has provision for the two discharge lines, which increases its versatility and field capacity. The plunger pump being a positive displacement pump, builds up a high pressure to throw spray liquid to larger distances with a suitable boom. The pump barrel, lance and the spray nozzle are made from brass alloy. For operation the inlet pipe is placed in the storage container and one person continuously operates the pump by foot lever. There is a provision for the operator to hold the sprayer at the top by U-type fixture. The other person directs the lance to the target. For spraying tall trees up to a height of 10 m, a high jet or bamboo lance can be used. The foot sprayer is all purpose sprayers, suitable for both small and large scale spraying on field crops, in orchards, vegetable gardens, tea and coffee plantations, rubber estates, flower crops, nurseries etc.

Fig. 7.5: Foot operated sprayer

Knapsack sprayer

Knapsack sprayer consists of a pump and a air chamber permanently installed in a 9 to 22.5 liters tank (Fig. 7.6). The handle of the pump extending over the shoulder or under the arm of operator makes it possible to pump with one hand and spray with the other. Uniform pressure can be maintained by keeping the pump in continuous operation. Knapsack sprayers are used for spraying insecticides and pesticides on small trees, shrubs and row crops.

Fig. 7.6: Knapsack sprayer

Power sprayer

Power sprayers are used for developing high pressure and high discharge for covering large area. These sprayers are either operated by auxiliary engines or electric motors. Most of these sprayers are hydraulic sprayers and consist of power unit to drive the pump, pump unit which employs piston or plunger pump, piston (1 to 3), pressure gauges, pressure regulators, air chamber, suction pipe with strainer, delivery pipes fitted with lance, gooseneck bend and nozzles (Fig. 7.7). The portable sprayers use petrol engine so that these can be easily taken to the spray sites. The complete assembly is mounted on the stretcher type frame or on wheel barrow for easy transportation. The number of lances may vary from 1 to 6 depending upon the model. In some models there is a built in storage tank of fibreglass having capacity of 100 litres, while in others a separate storage tank is required in which the suction pipe of the sprayer remains submerged. For operation, the shut off trigger valve of the lance is closed and the engine/ electric motor is started to actuate the pump. The pump draws the spray liquid from the tank, imparts pressure energy and sends it to the delivery line/lines. The operator directs the lance towards the target and operates the trigger/shut off valve. Adjusting the nozzle or selecting the appropriate nozzle, adjusts the spray pattern. For delivering the spray liquid to large distances/ height a bamboo lance can also be used. Field capacity is 0.2-0.3 ha/h and discharge rate up to 25 l/min. These sprayers are suitable for spraying in orchards, tea and coffee plantations, rubber plantations, vineyards and field crops. Tall trees up to a height of 15 meters can be sprayed with these types of sprayers.

Fig. 7.7: Power sprayer

Motorized knapsack sprayer

a) Motorized knapsack sprayer in use

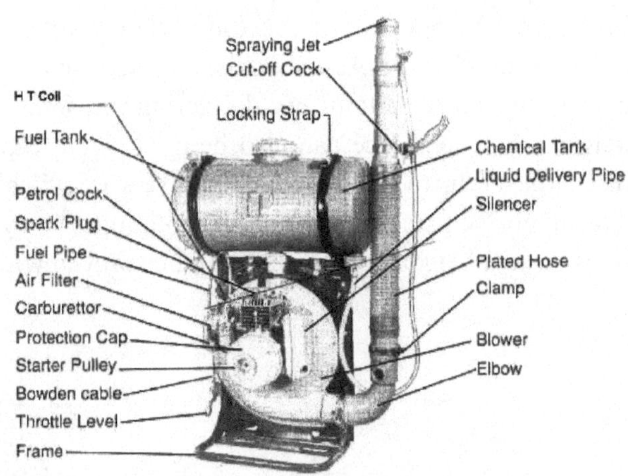

b) Details of motorized knapsack sprayer

Fig. 7.8: Motorized knapsack sprayer

The motorized knapsack mist blower has a small 2-stroke petrol/ kerosene engine of 35 cc to which a centrifugal fan is connected (Fig. 7.8). The centrifugal fan is usually mounted vertically. The fan produces a high velocity air stream, which is diverted through a 90-degree elbow to a flexible (plastic) discharge hose, which has a divergent outlet. The spray tank that has also a compartment for fuel and engine-fan unit is mounted on a common frame, which fits to the back of operator. The tank is made of plastic. The spray liquid flows due to gravity and suction created at the tip of nozzle, thus remains in the air stream. Some models of the sprayers have roller pump for pumping the spray liquid into the discharge hose and have tall tree spraying attachment. These sprayers have shear type nozzles. For operation, the tank is filled with the spray liquid and cranking with the rope starts the engine. Upon rotation of the engine the fan produces a high velocity air stream. The control valve for the spray liquid is opened gradually and adjusted for the desired flow rate. The operator directs the discharge hose to the target. The spray liquid that falls in the air stream gets sheared and upon coming in contact with the atmosphere a mist is produced. Air velocity at the outlet is 65-75 m/s and field capacity 3 ha/h. It is used for spraying in orchards, coffee estates and tall crops.

Tree sprayers

The tree sprayer is an ideal sprayer for spraying tall fruit trees in the orchards. It consists of a 4- stroke petrol/ kerosene engine to drive the fan, a centrifugal fan which produces air stream of high volume and velocity, a micronizer nozzle for producing uniform and fine droplets of spray liquid in the range of 150-200 microns, plastic tank for storage of spray liquid, rotary pump to draw the spray liquid from the tank and to feed it to the nozzle and a fibre glass casing (Fig. 7.9). All these components are joined and mounted on the stretcher type of frame. The sprayer can be carried by two persons to the place of spraying. For operation, the tank is filled with spray liquid and the engine is started by cranking with the rope, to drive the fan and the rotary pump. The control valve is opened to adjust the rate of flow. The sprayer is placed under the tree and manually moved around it to complete the spraying operation. Fan output is 1000 m^3/h and it can go up to 20-25 m height. Field capacity ranges between 3 and 4 ha/ day. The sprayer is used for spraying of tall fruit trees.

Fig. 7.9: Tree sprayer

Knapsack power sprayer

The pump adopts horizontal gear driving. It is powerful and stable in pressure. As it is oil-soaking type, the lubricating effect of crank box is positive. It adopts double cylinder pump, which raises operational efficiency (Fig. 7.10). The piston is heat treated and wear resistant. 'V' packing adopts special materials so it is durable. The engine has electronic ignition, which is easy to operate and maintain. The engine has high power to weight ratio. It uses gasoline as a fuel and produces one power stroke for each 180° of crank rotation. Pressure is controlled by oil valve; spraying pressure is changeable freely up to 30 kgf/cm^2. Its structure is rigid; materials are excellent and easy to be maintained. It is suitable for spraying pesticides and fungicides on rice, fruits and vegetable crops.

Fig. 7.10: Knapsack power sprayer

Battery operated low volume knapsack spinning disc sprayer

The sprayer consists of plastic tank, grooved spinning disc, a fractional horsepower DC motor to spin the disc at very high speed, 12 volt power supply source which can be a dry battery or lead acid battery and a light weight handle on which the spray head is mounted. The handle has provision to adjust the angle of the spray head. The spinning disc is also made of fine quality plastic, which has very fine radial grooves. For operation, the spray liquid in minimum dilution is filled in the tank, which flows in the form of drops at the centre of spinning disc. The disc is attached to the motor, which rotates it at a very high speed. The spray liquid travels in the grooves of the disc and gets fragmented into very fine droplets when it leaves the periphery of the disc. The droplets are thrown outwards due to centrifugal force. The tank capacity is of 10 litres and it can apply 45 litre/ha. The field capacity is 0.20 ha/h. It is suitable for spraying in crops like paddy, cotton, groundnut, pulses and vegetable. It saves 30 per cent labour and operating time and 15 per cent on cost of operation compared to manual spraying. It also results in 27 per cent increase in yield compared to spraying by manual sprayer

Hand rotary duster

The hand rotary duster is available in two models, shoulder mounted and belly mounted. It is a common type of duster being used by the farmers. The duster consists of a hopper, fan/blower, rigid/flexible discharge pipe, reduction gearbox, rotating handle, shoulder straps, and metering mechanism (Fig. 7.11). The hopper is either made of plastic or aluminium. The hopper made from mild steel sheet is coated with anti-corrosive material for longer life. The duster has mechanical agitator connected to the gearbox placed in the hopper, which churns the chemical and prevent clogging of the outlet. The adjustable orifice plate mounted below the hopper outlet controls the application rate. The fan/ blower is enclosed in the casing and is rotated with the handle through gearbox. For operation, the hopper is filled 1/2 to 3/ 4th of the capacity. This is mounted on the shoulder/belly with the help of adjustable straps. The discharge pipe fitted with spoon type deflector is directed towards the target continuously rotating the handle. The chemical in dust/powder form drops from the hopper in the discharge pipe having an air stream created by the blower. These dust particles emerging in the form of cloud from the discharge pipe are carried to the plant where these settle on the leaves, stems and other parts. The hopper capacity is of 5 litres and field capacity 0.6 ha/day. This is used for control of pests and diseases by use of chemicals in the dust forms in nursery, vegetable gardens, field crops, tea and coffee plantations, green houses, glasshouses and godowns.

Fig. 7.11: Hand rotary duster

Tree duster

Tree duster is used for dusting of the orchards and tall trees. The tree duster consists of an engine which may be petrol or petrol/ kerosene run, blower/ fan directly coupled to engine that produces a high volume and high velocity air stream, hopper for the storage of the chemical, discharge chute and other control attachments (Fig. 7.12). All the sub-assemblies are joined together and mounted on the stretcher frame. The equipment can be conveniently carried by two persons to the dusting place and moved around the periphery of the tree. For operation, the hopper is filled with the chemical and the metering mechanism is adjusted for the desired application rate. The equipment is placed under the tree to be dusted and cranking starts the engine. The chemical drops in the discharge chute where it comes in contact with high velocity air stream and is thus carried to the target. The air

Fig. 7.12: Tree duster

velocity imparts enough energy to the dust particles for their penetration in the canopy of the tree. The chemical usually comes out from the discharge chute in the form of cloud. The equipment is moved around the tree for completion of the dusting with the help of stretcher frame and moved to the basin of next tree. The air velocity produced with this duster is about 70 m/s. The duster can throw the powder to the height of about 25 m. The capacity of the duster is 3 to 4 ha/day depending upon the plant distance and density. The fan output is 15-20 m^3/min. It can be used for dusting in orchards and tall trees.

Power tiller mounted orchard sprayer

It consists of an HTP (horizontal triplex piston) pump, trailed type main chassis with transport wheels, chemical tank with hydraulic agitation system, cut off device and boom equipped with turbo nozzles (Fig. 7.13), Pandey *et al.,* 1997. It is fitted with turbo nozzles with operating pressure of 9-18 kg/cm^2. It generates droplets of 100-150 micron sizes. Depending upon the plant size and their row spacing, the orientation of booms can be adjusted. The spray booms are mounted behind the operator. The tank capacity is of 200 litres. It has six nozzles three mounted on each side. Nozzle spacing can be adjusted. The boom covers half of the tree canopy on either side of the sprayer. It is suitable for spraying chemicals in orchards like grapes, citrus, pomegranate etc. It can cover 0.40-0.70 ha/h at travel speed of 1.20-1.50 km/h.

Fig. 7.13: Power tiller mounted orchard sprayer

Front mounted self-propelled boom sprayer

It consists of chassis with ground drive system, diesel engine (5.5 hp), spray boom with hollow cone nozzles, power transmission system, gear box, lugged cage wheels, chemical spray tank and pump (Fig. 7.14), Pandey *et al.*, (1997). The pump and boom with 14 nozzles are mounted on a stand, which can be attached in front of the frame. It can develop nozzle pressure of 3 kg/ cm^2. It can cover 0.60 ha/h with application rate of 500 litres per hectare. It can cover 6.5 m width. It is used for uniform spraying in horticultural crops and other crops like cotton, maize, and groundnut.

Fig. 7.14: Front mounted self-propelled boom sprayer

Self-propelled light weight boom sprayer

Fig. 7.15: Self-propelled light weight boom sprayer

The machine is operated by 5 hp diesel engine and is controlled by the operator from the handle. The machine consists of a lightweight power tiller unit and a spraying unit. Spray pump and two narrow pneumatic wheels get power from the engine through gears, chains and sprockets (Fig. 7.15), Garg and Singh (2002), Singh and Pandey (2008), Anonymous (2012). Spray pump is of Roller type. One caster wheel is also provided at the rear, which acts as supporting wheel. The spray boom is mounted on the power unit through a canopy frame. To meet the requirement of a high capacity effective sprayer, a self-propelled boom sprayer with 14 nozzles mounted on a frame of self-propelled vertical conveyor reaper. The spacing of nozzles can be varied from 300 to 600 mm. Height of boom can be varied from 60 to 130 cm. Nozzles used are of generally flat fan type but any other type can also be used. Ground clearance is about 50 cm. Tank capacity is about 100 litres. It can cover 3-5 ha/day with width of coverage of 6.3 m. The self-propelled lightweight boom sprayer is used for chemical application on wheat, vegetable and other crops

Tractor mounted sprayer

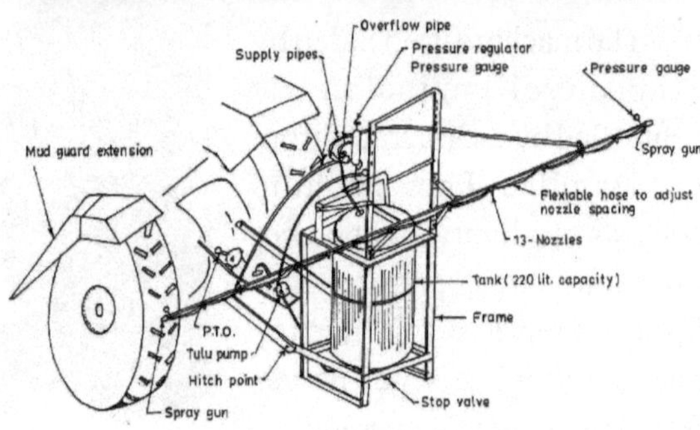

Fig. 7.16: Tractor mounted sprayer

These are hydraulic energy sprayers. They utilize PTO power of the tractor to operate the pump of the sprayer. Basically the spray boom can be arranged in two ways; ground spray boom and overhead spray boom. The overhead spray boom is designed for tall field crops and the planting is done in such a way that it leaves an unplanted strip of about 2.5 m width for operation of the tractor. Therefore a planted strip may be 18-20 m wide and after every planted strip a fallow strip has to be left for tractor operation. For ground spray boom the planting has to be done in rows keeping in view track width of the tractor. It is suitable for use when the crop is small. The sprayer essentially consists of a tank which is made of fibre glass or plastic, pump assembly, suction pipe with strainer, pressure gauges, pressure regulators, air chamber, delivery pipe, spray boom fitted with nozzles of different types (Fig. 7.16), Pandey *et al*., (1997), Shukla *et al*., 1987. The complete sprayer is mounted on 3-point linkages of the tractor. It uses high pressure and high discharge pump as the number of nozzles may be up to 20 depending upon the crop and make of the sprayer. Field capacity is about 8 ha/day with 14 nozzles. Number of nozzles will depend on boom length. It is used for spraying in vegetable gardens, flower crops, vineyards and for tall field crops like sugarcane, maize, cotton, sorghum, millets etc.

Tractor operated blower sprayer

These are tractor mounted or trailed air carrier type of sprayers for spraying in orchards. The spray volume can be controlled making it high volume, low volume or ultra low volume sprayer depending upon the requirement. The sprayer mainly consists of high-pressure piston pump, which atomize the spray solution, axial or centrifugal fan to produce a stream of air, interchangeable and orientable spray heads, high capacity spray tank and other control attachments (Fig. 7.17), Pandey *et al.*, (1997). All the sub-assemblies are mounted on the frame and can be attached to 3-point linkage of the tractor or mounted on a trailer. The spray heads are mounted radially so as to cover the 2-rows of the trees. The spray heads can be oriented upwards for spraying the tall trees. In case of spraying individual tree the spray heads can be exactly adjusted to the form of tree. The volume of spray and the air stream velocity and volume can also be adjusted as per the requirement. For operation the spray liquid is filled in the tank. The tractor power is used to drive the fan and pump. The pump sucks and sends the liquid to the spray head, which atomize the liquid, and at the same time droplets meet the air stream to carry them to the target. These sprayers can be used for chemical spraying in tall trees, plantation crops, vineyards and other field crops.

Fig. 7.17: Tractor operated blower sprayer

Self-propelled high clearance sprayer

The machine has a chassis with 1200-mm ground clearance, four wheels, and 20 hp diesel engine, gearbox, water tank (220 litre capacity), seat for the operator, spray pump and boom with18 nozzles (Fig. 7.18), Garg and Singh (2002), Anonymous (2010, 2012). The nozzles are spaced at 675 mm and total boom width is 10.80 m. The boom height can be adjusted form 31.5 cm to 168.5 cm to suit different crops and can be folded during transport. The front two wheels are narrow in width (20 cm) and are given drive, while the rear wheels are steering wheels. Fenders have been provided in front of the drive wheels to deflect the crop branches away from the wheels for reducing mechanical damage. The wheel track is 135 cm and during operation 2 rows of cotton crop come under the machine chassis. Machine has four forward speeds and one reverse speed. The 1st and 2nd gears are for field operation while 3rd and 4th gears are for road transport. The field speed is up to 5 km/h and the road speed is up to 25 km/h. The effective field capacity varies from 1.6 to 2.0 ha/h. The average discharge from the nozzle is 1.4 l/min at 3.5 kg/cm^2. Self-propelled high clearance sprayer is most suitable for spraying on tall crop like cotton. The machine can also be used for spraying on sunflower, wheat and other crops.

a) Self-propelled high clearance sprayer in use in field

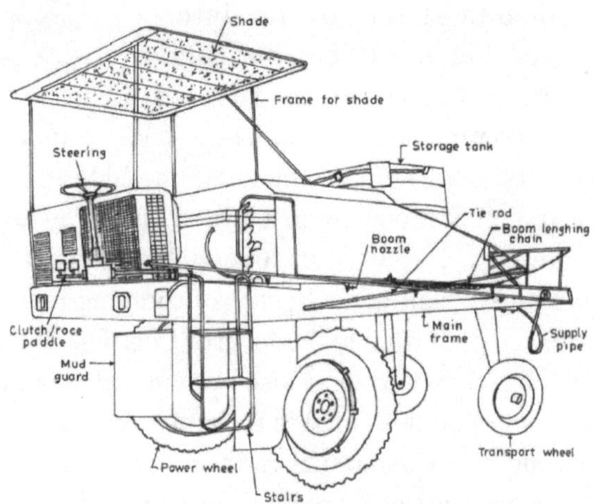

b) Details of self-propelled high clearance sprayer

Fig. 7.18: Self-propelled high clearance sprayer

Tractor operated aero blast sprayer

The farm machine has great potential for its introduction in the cotton crop growing area, because as many as 6-7 sprays are required for complete control of insects. Tractor operated sprayers damage the crop, which comes under chassis especially when the crop is tall. On the other hand, aero-blast sprayer will result in low crop damage and hence contribute towards increase in yield. Also, by introduction of this sprayer the farmer's having large orchards may be benefited. Tractor operated aero blast sprayer can be used for spraying of chemical in orchards of mango, guava, sapota, jack fruits, ber, lemon and similar height orchard plantations and tall crops like cotton, pigeon pea and sunflower. The area under these crops may also increase with introduction of this machine. Because of high capacity machine, this may also promote custom hiring. The machine consists of a tank of 400 litres capacity, pump, fan, control value, filling unit, spout adjustable handle and spraying nozzles to release the pesticide solution in to stream of air blast produced by the centrifugal blower (Fig. 7.19), Anonymous (2008, 2010, 2012), Singh and Pandey (2008). The air blast distributes the chemical in the form of very fine particles throughout its swath width, which is on one side of tractor. The major portion of swath width is taken care of by the main blast through the main spout and the supplement nozzles cover the swath area near the tractor. The sprayer is mounted on the tractor 3-point linkage and is operated by tractor PTO. Application rate of sprayer can be varied from 100-400 litres/ha depending upon different valve setting. Effective width of sprayer is found to be about 13.0 m distance. The sprayer consists of one blower having air discharge of 1–3 m^3/s, one centrifugal pump/triplex pump, flow control valves and cone type rotary nozzles. The discharge of the nozzle varies from 3.8 to 6.4 l/min when one bypass valve is closed and 11 to 19 l/min when both valves are open at PTO speed of 300 to 540 rpm. The effective field capacity of the sprayer is about 1.5 ha/h at the forward speed of 3.5 km/h. It helps to increase yield by way of better insect & pest control. Better control of insects, human drudgery is greatly reduced in comparison to knapsack sprayers, minimizes chemical contact with humans/user, cost effectiveness not much but quality of work is better.

a) Aero blast sprayer in use

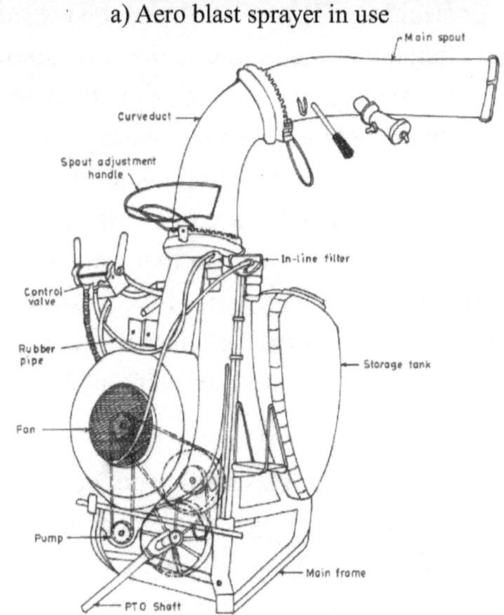

b) Details of aero blast sprayer

Fig. 7.19: A view of tractor operated aero blast sprayer.

Tractor operated air assisted horizontal boom sprayer

Most of the Indian farmers use knapsack sprayers or tractor mounted hydraulic sprayers for spraying on different crops. These hydraulic sprayers apply pesticides on upper canopy of the plants. The upper canopy of the plant foliage prevents sprayed droplets from reaching the lower leaves and especially from reaching underside of leaves. Pests such as aphids, white flies etc usually feed on the underside of the leaves and down into the plant canopy. Therefore, an efficient spray technology is most important in enhancing the effectiveness of pesticides. Optimum droplet size, droplet density and their uniform impingement to the target surface are needed to ensure an efficient pest control. Air-assisting systems are designed to increase effectiveness of pest-control substances, provide better coverage to underside of leaves, promote deeper penetration into crop canopy, and make it easier for small droplets to deposit on target, cover more area per load and reduce drift. Air stream also creates turbulence within the crop that improves deposition of spray material on the targets, which are normally inaccessible. However to reduce the frequency of application, quantity of chemical and to obtain better coverage of chemicals on cotton crop, air-assisted sprayer attachment to tractors should become available to farmers.

An air-assisting system produces a continuous airflow, which consists of an axial fan, its casing and ducts (Fig. 7.20), Anonymous (2010), Anonymous (2012), Singh and Pandey (2008). Axial fan is designed and fabricated to provide air at a required flow rate and pressure. Size of the impeller is 73cm having hub of 32 cm and nine aluminum blades of 20.5cm length. The angle of each blade at the base is kept 22 degrees and at the tip 6 degrees. There is provision to change the angle of the blade. The fan casing is made out of G.I. sheet. Fan casing is kept open from the top and on the bottom of the casing there were sideways circular air outlets of 46 cm diameter. On the air outlets of the fan casing, sleeves (ducts) of 46 cm diameter are fixed. Throughout the length of the sleeve (duct) 40 mm holes 80 mm apart are made on bottom side. The total length of the sleeves (ducts) plus casing width is 6 m. A horizontal triplex pump (HTP) is used to generate pressure for the application of spray formulation. A boom of GI pipe having diameter of 35 mm is used for nozzle mounting. Spacing between nozzles, its working pressure and height are selected as 45 cm, 3.5 kg/cm^2 and 50 cm respectively for uniform spray distribution. Nozzle boom is placed just bellow but slightly behind of the holes of the sleeves. A 35 hp tractor is used to operate the sprayer. The uniformity coefficient for air-assisted sprayer is better (1.69) than for conventional sprayer (2.04). There is sufficient droplet deposition on the upper side of leaves of any section of plant for both the sprayers. Due to air-assistance, the sprayer is able to put an effective number of drops per square centimeter on underside of leaves at any section of plants. The reduction in number of whitefly is 29.5 to 69.7% more for air-assisted sprayer as compared to conventional sprayer. The reduction in freshly shed damaged fruiting bodies is 30 to 38.33% for air-assisted sprayer where as for conventional sprayer it is 16 to 23.33%.

a) A view of Tractor operated air assisted horizontal boom sprayer
(*Courtesy*: Gurusukh Agro Works, Samrala)

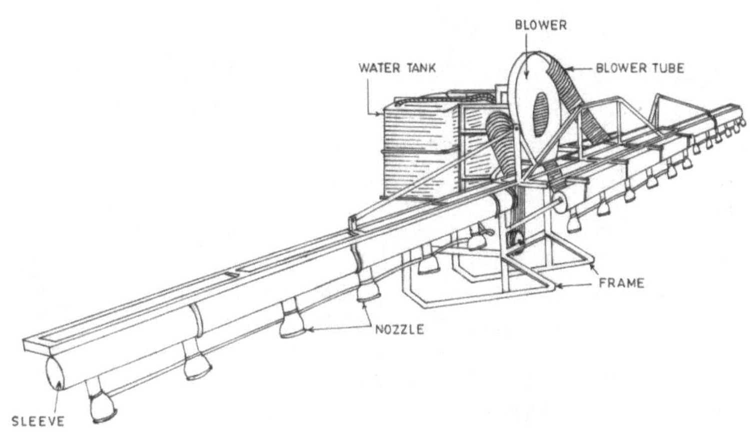

b) Details of tractor operated air assisted horizontal boom sprayer
Fig. 7.20: Tractor operated air assisted horizontal boom sprayer

Tractor operated air assisted vertical boom sprayer

Tractor operated air assisted vertical boom sprayer is similar to tractor operated air assisted horizontal boom sprayer. The only difference is that air assisted boom is in vertical position in this case (Fig. 7.21). Air assisted vertical boom sprayer can be used effectively to spray the insecticide and other chemicals to the fruit crops such as orange, lemon, guava, sapota etc due to air blast created. One side each of two rows of trees can be covered in one go as shown in Fig. 7.21. It can cover about 1.7 ha/h with more than 80% field efficiency.

Fig. 7.21: Air assisted boom sprayer in vertical position.
Courtesy: Gurusukh Agro Works, Samrala

Nozzles

Nozzle is an important mechanism, which breaks the spray liquid into the desired size of droplets for application to the surface to be sprayed. The nozzles help to control the rate, uniformity, thoroughness and safety of pesticide application. The principal parts of nozzles are body, valve, washer, disc and cap. The shut-off may be provided in the supply line or in the nozzle itself. The valve is used to impart the whirling motion to the spray liquid as it flows at high velocity through valve into eddy chamber of nozzle body. The whirl mechanism is usually a spiral on the valve that fits into nozzle body. Since no single nozzle can meet various spray requirements, they are now commonly manufactured with inexpensive replaceable nozzle tips or discs, which can be selected to give, desired spray characteristics and volume for the specified job. Nozzles vary with respect to rate of discharge, the angle of spray and the type of spray pattern. Keeping these characteristics into mind, five types of nozzles commonly used in pesticide spray equipment are hollow-cone, solid-cone, triple-action, flat fan and flooding nozzles (Fig. 7.22). The hole at centre of nozzle disc

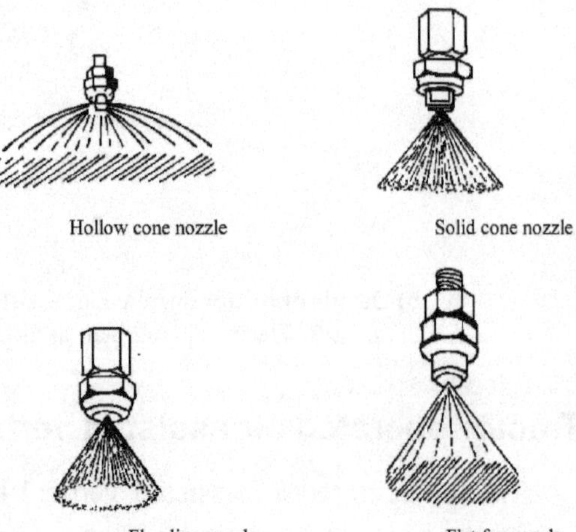

Fig. 7.22: Different types of nozzles used in sprayers.

results in a hollow cone of spray. If there is a hole through the centre of valve, a solid cone of spray is produced. Built into some nozzles is a removable strainer with slightly smaller openings than the nozzle orifice to prevent clogging. Most of these nozzles require a minimum hydraulic pressure of 75 kPa whereas for full development of spray pattern a hydraulic pressure of 300 kPa is required.

Hollow-cone nozzle: They are commonly used for spraying insecticides and fungicides. The pattern of spray is circular with tapered edges to form a hollow cone or ring type spray pattern. This type of nozzle provides adequate coverage of pesticides from all angles. The liquid is fed into a whirl chamber through tangential entry or through a fixed spiral passage to give a rotating motion. The liquid comes out in the form of a hollow conical sheet, which then breaks into small drops.

Solid-cone nozzle: These nozzles produce a solid-cone spray pattern and are used for spot spraying of herbicides in lawns. They cover entire circular area evenly. The construction is similar to hollow cone nozzle with the addition of an internal jet, which strikes the rotating liquid just within the orifice of discharge. The breaking of spray droplet is mainly due to impact action.

Flat-fan nozzle: The flat-fan nozzle produces a spray pattern in the form of a flat sheet and is used for spraying herbicides. The spray pattern has been designed in such a way that 30-50 per cent overlap can be achieved for even distribution of spray liquids. The nozzle tip is provided with an outlet that is elliptical. This nozzle is mostly used for low pressure spraying.

Flooding nozzle: It makes a wide-angle flat spray pattern. It works at lower pressure than other nozzles. The spray pattern is fairly uniform across its width. It is used for spraying herbicides and fertilizer solutions. The flooding flat nozzles result in less drift than the other type of nozzles.

Triple-action nozzle: These nozzles are very popular because they can produce a jet, a solid-cone and a hollow-cone, three types of spray pattern by adjustment. They are used for high-volume as well as low-volume spray applications and are mostly used with tractor-drawn sprayers. The triple action nozzle consists of a swirl plate, swirl chamber, a nozzle disc and a cap. Nozzles are usually made of brass, stainless steel or plastic. Pressure has a considerable effect on nozzle performance. For a given size and design of nozzle and depth of swirl chamber, the higher the pressure, the higher the throughput, the wider the spray angle and the finer the spray droplets.

Selection of nozzles

Most of the nozzle manufacturers give discharge of various nozzles manufactured by them at different pressures. This information can be used for selection of correct nozzle for spray job to be done. Before selecting the right nozzle, one should consider the points such as i) type of spray job i.e. spraying of weedicides, insecticides etc, ii) total amount of spray solution to be applied, iii) row spacing, iv) number of nozzles to be used per row and nozzle spacing, v) type of spray pattern desired i.e. fan or cone type, vi) speed of travel, and vii) approximate pressure to be used in spraying.

It has been observed that the use of single hollow cone nozzle directly over the crop row gives better insect control and results in higher yields. The fan and boomless- type spray nozzles do not give as good results as hollow nozzles. The higher yields are obtained from the rows nearest to the nozzle and where more complete coverage is achieved with smaller droplets. Proper nozzle adjustment and care are required to maintain good nozzle performance. The distance of nozzle tips from the sprayed surface, the angle of spray and the spacing of nozzles all have a bearing on proper spray coverage. Nozzles must be spaced equally along the spray boom and positioned correctly in relation to the row. The most common problems with nozzles are: i) worn out or plugged nozzles that result in untreated strips, ii) misaligned nozzles that result in reduced coverage, iii) improper nozzle tips resulting in irregular spray coverage, iv) spray boom not being level resulting in irregular coverage, and spray boom being too low or too high resulting in uneven patterns and coverage.

Maintenance of pesticides application equipment

Pesticides are expensive and so they must be applied efficiently. Proper maintenance of spraying equipment is, therefore, essential. Following suggestions would help minimize the labour problems and prolong the useful life of pesticide application equipment:

- Use clean water.
- Keep screens in place.
- Never use a metal object to clear the nozzles.
- Flush sprayers before using them.
- Clean sprayers thoroughly after use.

- Remove and clean all screens and nozzle tips in kerosene using a soft brush.
- When pump is not in use, fill it with a light oil and store it in a dry place.
- Lubricate all the points required.

Safety precautions

All pesticides are poisonous to human beings and hence enough safety precautions are required to be taken. Remember the following points:

- Always keep pesticides under lock and key.
- Do not store leaky containers.
- Do not keep pesticides near the food and fodder storage.
- Read the label and directions for use before opening a pesticide container.
- Do not use cooking utensils for preparing the solutions.
- Ensure that sprayer is not leaking.
- Always use safety wares like apron, goggles, facemask, gumboots and gloves while spraying.
- Remove nozzle clogging by thin pin and not by mouth.
- Smoking, eating, drinking etc during pesticide application is strictly prohibited without washing hands, face, body etc.
- Do not store spray solution over night.
- Clean the spraying equipment thoroughly to decontaminate it.
- Change the clothes and wash them after spray.
- Have a thorough bath with soap and water.
- Store the unused contents of container in the original container.

Literature Cited

Anonymous. 2008. Research Highlight. AICRP on Farm Implements and Machinery, CIAE Bhopal. Technical Bulletin No.: CIAE/2008/141.

Anonymous. 2010. Research Highlight. AICRP on Farm Implements and Machinery, CIAE Bhopal. Technical Bulletin No.: CIAE/2010/151.

Anonymous. 2012. Directory of Successful Farm Machinery in SAARC Countries. SAARC Agriculture Centre. BARC Complex, Farmgate, Dhaka – 1215 (Bangladesh).

Garg I K; Singh Surendra. 2002. Farm equipment for Punjab agriculture. Department of Farm Power & Machinery, Punjab Agricultural University, Ludhiana.

Pandey M M; Majumdar K L; Singh Gyanendra; Singh Gajendra. 1997. Farm Machinery Research Digest. Technical Bulletin No. CIAE/97/69, Central Institute of Agricultural Engineering, Bhopal, 328 p.

Shukla L N; Sandhar N S; Singh Surendra; Singh Joginder. 1987. Development and evaluation of wide-swath, tractor-mounted sprayer for cotton crops. Agricultural Mechanization in Asia, Africa and Latin America (AMA) Vol. 18 (2): 33-36.

Singh Surendra. 2007. Farm Machinery – Principles and Applications. Directorate of Information & Publication of Agriculture, Indian Council of Agricultural Research, Krishi Anusandhan Bhawan-I, Pusa Campus, New Delhi.

Singh Surendra; Pandey M M. 2008. X Plan Achievements (2002-2007). AICRP on Farm Implements and Machinery, CIAE Bhopal. Technical Bulletin No.: CIAE/2008/137.

Singh Surendra; Kandoria J L. 1999. Selection, Use and Maintenance of Pesticide Application Equipment. Directorate of Information & Publication of Agriculture, Indian Council of Agricultural Research, Krishi Anusandhan Bhawan-I, Pusa Campus, New Delhi.

8 Harvesting and Threshing Equipment

Harvesting and threshing are the key most important operations in the entire range of field operations required in any crop production system. Both of these operations remain still the most important operations as fruit of labour depends upon them. Crops are harvested after normal maturity with the objective to take out grain, straw, tubers etc without much loss. It involves cutting / digging/ picking, laying, gathering, curing, transport and stacking of the crop. Harvesting of major cereals, pulse and oilseed crops are done by using sickle whereas tuber crops are harvested by country plough or spade (Fig. 8.1). In case of cereals like wheat and paddy the plants are straight and smooth and ears containing grains are at the top whereas most of oilseed and pulse crops have branches which create problems in harvesting by manual or mechanical means. As per Bureau of Indian Standards the cutting and conveying losses should not be more than 2 per cent. Timeliness of harvest is of prime importance. Rapid harvest facilitates extra days for land preparation and earlier planting of the next crop. The use of machines can help to harvest at proper stage of crop maturity and reduce drudgery and operational time. Considering these, improved harvesting tools, equipment, combines are being accepted by the farmers.

Fig. 8.1: Manual harvesting of rice crop by sickle

Harvesting of wheat and rice is traditionally done by using local sickle (Singh, 2007; Singh and Verma, 2009; Pandey *et al.*, 1997).

Improved serrated blade sickles are also in use. Sorghum is harvested by local sickle. Suitable machines are not available for harvesting this crop. However, combine harvesters are in use in advanced countries. For maize the traditional practice is to collect the matured cobs manually. Grain combines equipped with corn-head snapping unit are being used in developed countries. Pulse crops such as Bengal gram, Pigeon pea, Urad, Moong and Cowpea are harvested by local sickle. Improved serrated blade sickles are also in use. The performance of narrow pitch cutter bar with horizontal conveyor is better than other types of available reapers. Combines with floating cutter-bar are in use in advanced countries. Oilseed crops such as Rapeseed and Mustard is harvested manually using sickles. In tall varieties, farmers cut the plants above ground level and leave long stubbles in field, which are subsequently ploughed in. In some areas, where plants are used as fuel or thatch material, harvesting with serrated blade sickles close to ground level is practiced by farmers. Mechanical harvesting of crop is also carried out using self propelled walk behind type vertical conveyor reaper, tractor front or rear mounted PTO operated reaper, and combine harvester. Combine harvesters need slight modification to make them suitable for harvesting rapeseed crop because of tall and branchy nature of the crop. The changes required are (a) attachment of vertical cutter bar at outer ends to get a clean cut windrow and (b) increase of header table length to accommodate the bulky crop. Combine harvesters can handle the crop easily after above changes are made.

Digging of groundnut crop with country plough and blade hoe at proper soil moisture level and manual pulling and gathering of pods using hand hoe is common practice. Animal drawn and tractor operated diggers and digger windrowers are improved implements developed for groundnut harvesting. The blade harrow is widely used for digging of groundnut crop in Gujarat. Tractor operated groundnut diggers have wide blade, which cover 1.25 to 2 m width and operate at 10 to 15 cm depth. Soybean harvested by local sickle is the traditional practice followed by farmers. However, modified serrated blade sickles are recommended as plant stem is 8 to 12 mm thick. Self-propelled vertical conveyor reaper windrower, tractor rear mounted reaper and combine harvesters are also used for soybean harvesting. When the available harvesters are to be used for soybean, these are required to be modified and adjusted to reduce field losses and suit crop and soil conditions. Cutting of crop close to ground with low stubble height and crowding and stripping effect are the main requirements.

Combine harvesters with floating cutter bars are recommended for low harvesting losses. Narrow pitch cutter bar has been reported to give lower harvesting losses as compared to conventional cutter bar. The traditional practice is to manually harvest the flower heads of sunflower and castor plants. These are stacked and sun dried for threshing. Suitable machines are not available for harvesting of sunflower and castor crops. Harvesting of whole plant would require separation of flower heads for threshing and thus the time saved by harvesting the whole plants would not reduce the labour requirement. The combine harvesters are used for harvesting of above crops using specially designed header. These are in use in advanced countries. The traditional practice to harvest safflower is to manually harvest the crop using sickles. Because of thorny and spiny nature of crop, harvesting and handling of safflower plants is a problem. Use of hand gloves and covers on legs and arms is recommended during harvesting. Hay forks are used for gathering and stacking the plants in field or in trailers. For mechanical harvesting of safflower self-propelled (1 m wide) vertical conveyor reaper and combine harvester are used. For harvesting tall varieties, there are problems as plants in rows are entangled with each other. Therefore, in combine harvesters, a vertical cutter bar is used at outer end to cut and separate the plants of harvested row. Similar cutting device is also used on reapers on the outer crop row divider and the belt conveyor has to be raised up, to take care of tall crop. Sesamum and linseed are harvested manually using sickles. Tractor rear and side mounted reaper can be used for harvesting the broadcasted crop. Vertical conveyor reapers have been used for harvesting crop, raised in rows and at optimum moisture level, i.e. 15-20 per cent, to avoid shattering of pods.

Factors affecting performance of harvesting machines

Crop factors: Crop variety, ambient temperature, maturity of crop, crop moisture, crop condition and crop density.

Machine factors: Shape and size of crop divider, reel position and speed, cutting blade shape and speed, conveyor speed, machine vibrations, and machine settings.

Operational factors: Height of cut and operational speed

Different type of mechanical harvesting tools / equipment, suitability for crops and their limitations

Serrated blade sickle: It has a plain/serrated curved blade and a wooden handle (Fig. 8.2). The handle of improved sickle has a bend at the rear for better grip and to avoid hand injury during operation. Serrated blade sickles cut the crop by principle of friction cutting like in saw blade. The crop is held in one hand and the sickle is pulled along an arc for cutting. Cutting of crop close to the ground is possible with modified handle. Energy requirement is 80-110 man-h/ha. It can be used effectively for harvesting of wheat, rice and grasses. There are different types of sickles used in different parts of countries but most popular are Naveen sickle, Punjab sickle and Vaibhav sickle. It saves 25-30 per cent labour and operating time and 25-35 per cent on cost of operation compared to harvesting by local sickle. Field capacity of sickle is 0.018 ha/h and labour requirement 80 man-h/ha.

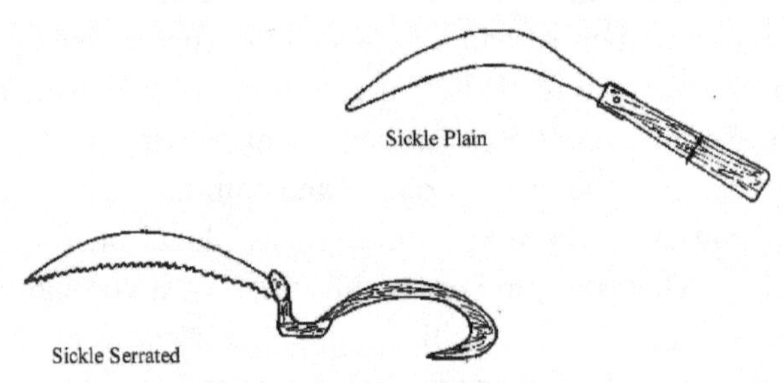

Fig. 8.2: Sickles used in harvesting of crops.

Reapers are used for harvesting of crops mostly at ground level. It consists of crop-row-divider, cutter bar assembly, feeding and conveying devices. Reapers are classified on the basis of conveying of crops as given below:

Vertical conveying reaper windrower: It consists of crop row divider, star wheel, cutter bar, and a pair of lugged canvas conveyor belts (Fig. 8.3). This type of machines cut the crops and conveys vertically to one end and windrows the crops on the ground uniformly. Collection of crop for making bundles is

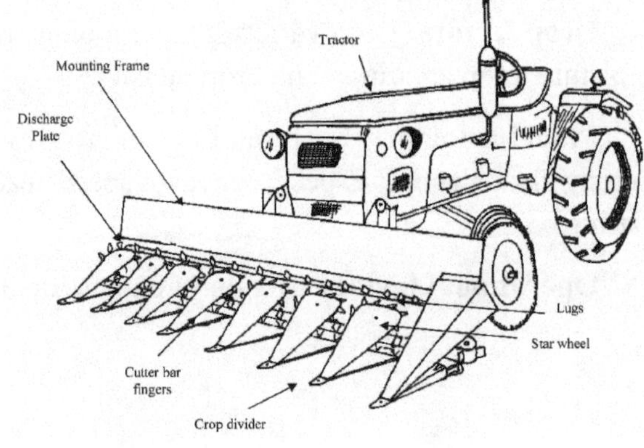

Fig. 8.3: Vertical conveyer reaper windrower

easy and it is done manually. Self-propelled walking type, self-propelled riding type and tractor mounted type reaper-windrowers are available. These types of reapers are suitable for crops like wheat and rice. The field capacities of these machines vary from 0.20-0.40 ha/h.

Horizontal conveying reapers: This type of reapers is provided with crop dividers at the end, crop gathering reel, cutter bar and horizontal conveyor belt. They cut the crop, convey the crop horizontally to one end and drop it to the ground in head-tail fashion. Collection of crop for making bundles is difficult. This type of reapers is tractor mounted and suitable for wheat, rice, soybean, and gram.

Bunch conveying reapers: This type of reapers are similar to horizontal conveying reapers except that the cut crop is collected on a platform and is being released occasionally to the ground in the form of a bunch by actuating a hand lever. Here, collection of crops for making bundles is difficult. Bullock drawn and tractor-operated models are available and they are suitable for harvesting wheat, rice and soybean crops.

Reaper binders: The cutting unit of this type of reapers may be disc type or cutter bar type. After cutting, the crop is conveyed vertically to the binding mechanism and released to the ground in the form of bundles. Self-propelled walking type models are available but these are not popular due to high cost of twine. Reaper binders are suitable for rice and wheat.

Strippers: The design of a tractor front mounted stripper is available for collection of matured grass seeds from the seed crops. It consists of a reel having helical rubber bats, which beat the grass over a sweeping surface where the ripened seeds get detached and the seeds are collected in the seed box.

Diggers: The design of groundnut and potato diggers of animal drawn and tractor operated types are available. The digging units consist of V-shaped or straight blade and lifter rods are attached behind the share. These lifter rods are spaced to allow the clods and residual material to drop while operating the implement. The plant along with pods/tubers is collected manually.

Combines: Various designs of combine harvester having 2 to 6 m long cutter bar are commercially available. The function of a combine harvester is to cut, thresh, winnow and clean grain/seed. It consists of header unit, threshing unit, separation unit, cleaning unit and grain collection unit. The function of the header is to cut and gather the crop and deliver it to the threshing cylinder. The reel pushes the straw back on to the platform while the cutter bar cuts it.

The crops are threshed between cylinder and concave due to impact and rubbing action. The threshed material is shaken and tossed back by the straw rack so that the grain moves and falls through the openings in the rack onto the cleaning shoe while the straw is discharged at the rear. The cleaning mechanism consists of two sieves and a fan. The grain is conveyed with a conveyor and collected in a grain tank.

Requirements of field and crops for harvesting by mechanical devices: The criteria met before mechanical harvesting equipment can be used successfully are: i) field must be fairly level without undulations to facilitate smooth operation and uniform stubble length, ii) for reapers and binders, plants must be grown in rows, iii) field efficiency of harvesting machines is high in large fields, and iv) water control in rice field is essential to ensure that the fields are drained and are relatively dry at harvest time.

Reapers

Harvesting of cereal crops especially wheat is a serious problem. There is a tremendous crop loss when untimely rain is experienced. Delayed harvesting causes grain shattering due to over maturity. The standing crop in the field can be harvested with the use of reapers. A reaper may be classified as animal-drawn reaper, animal-drawn engine operated reaper, tractor rear mounted PTO operated reaper, power tiller operated or tractor front mounted vertical conveyer type reapers and self-propelled reaper binder (Pandey *et al.*, 1997; Singh, 2007).

Animal-drawn reaper

It consists of a cutter bar of 1.05 m length. The power to drive the knife bar is given from the ground wheel by means of gearbox, crank and connecting rod mechanism. As the machine is pulled forward by a pair of bullocks, a reciprocating motion is imparted to the knife bar with a peak cutting velocity of about 100 m/min. The crop is cut due to shearing action. The effective field capacity of machine varies between 0.2-0.3 ha/h.

Animal-drawn engine operated reaper

It has a cutter bar of 1.35 m length. A 2-hp 4-stroke petrol engine operates the cutter bar. The drive to the cutter bar from engine is given through V belt and gearbox. It can cover 0.2-0.4 ha/h with field efficiency of about 80%.

Tractor-rear mounted side-operated reaper

It is suitable for harvesting of wheat, paddy and soybean crops. It comprises of a frame, 1.5 m wide cutter bar assembly, platform for collecting the crop and power transmission system to provide the drive to the cutter bar (Fig. 8.4), Garg and Singh (2002). The collected crop on the platform in the form of loose bundles of desired size is discharged at the back of the platform by controlling the platform level manually. The crop can also be delivered at the back of the tractor automatically by providing self raking mechanism or apron system. Machine is mounted on 3-point linkage of a 35-hp tractor and it gets the drive from the tractor PTO shaft. Since, the machine is rear mounted, so a swath of about 1.5 m width is required to be first cut manually all around the field to make passage for the tractor to travel. The machine is operated at a forward speed of about 3.0 km/h and can harvest 2-3 ha/day with the help of 3-4 persons. The harvested crop falling on the ground in the form of loose bundles can be tied into bundles of required size by the labour manually. Grain loss due to cutting and handling of crop is about 2 percent and at turns it ranged from 3 to 4 percent. The loss is reduced if the reapers have mechanical side delivery of the material. In this case manual shifting of the crop is not required to clear the passage for the subsequent run of the tractor. Height of cut can be adjusted with the help of tractor hydraulic and machine is raised while negotiating bunds. It saves 60-70% labour and 25-30% in cost of operation.

Fig. 8.4: Tractor-rear mounted side-operated reaper

Tractor-rear mounted reaper binder

The machine consists of a cutting, gathering, knotting mechanism mounted on a high pressure pipe frame with a 3-point linkage arrangement for hitching at the rear of a tractor. It has a 1.36 m long cutter bar and power to various components is given from PTO through v-belts and pulleys. The machine can cover 1.5-2.0 ha/day at a forward speed of 2 km/h. Machine can be used for harvesting wheat and paddy both. Grain losses are 2.2-8.0% for wheat and 1.0-5.0% for paddy.

Tractor front mounted vertical conveyer reaper windrower

The vertical conveyor reaper windrower has certain features, which enable it to have an edge over other designs of harvesters (Singh and Pandey, 2008; Singh, 2007; Anonymous, 2008, 2010, 2012). The main parts of the vertical conveyer reaper include star wheels, crop row dividers, cutter bar having knife sections of 76.2 mm pitch, lugged conveyor belts and power transmission system (Fig. 8.5). The effective cutter bar width is 2.1 m. Cutter bar is given reciprocating motion by crank wheel. The crop row dividers guide the standing crop, the star wheels direct the crop towards cutter bar and help in slightly lifting the crop after it is cut; turn it at 90^0 and help in conveying by the lugged conveyor belts. Two lugged flat belts convey the cut crops towards the right side of the machine keeping the crop in vertical orientation. At the end, the crop is discharged and laid on the ground in the form of a windrow. The power is transmitted from the power source to shaft through a long propeller shaft, which in turn drives the lugged flat conveyor belts and a crank to operate the cutter bar through a connecting rod. The top conveyor belt drives the star wheels on the crop row dividers. Pressure springs are fitted underneath the star wheels and to keep the cut crop in upright position while it is conveyed out of the machine. Vertically held crop is then delivered towards right side of the machine in a windrow perpendicular to the direction of movement of machine with the help of lugged conveyor belt. The gearbox and windrower is coupled to the drive shaft of the prime mower. The output shaft transmits power to the shaft driving the lugged flat conveyor belts and a crank is attached at the lower end of the output shaft to operate the cutter bar through connecting rod. The serrated blade cutter bar with standard knife guards is fitted. In tractor mounted models, the power to the gearbox is transmitted from PTO shaft through gearbox by a long shaft running beneath the tractor body to the front and with the help of universal joint and telescopic shaft which is connected to the gearbox. Lowering and raising of reaper is carried out with the help of hydraulic system of a tractor. The power in case of power tiller units is transmitted through an intermediate shaft to the gearbox on windrower either by belt or by shaft drive. In this case the machine pivots on the power tiller wheels. By pushing the handle, the cutter bar can be raised. The reaper windrower is lowered or raised by the hydraulic system of the tractor. The machine is mounted in the front of a tractor and is used for harvesting cereal crops like wheat, paddy, barley, soybean, rapeseed etc. The skilled labour requirement for harvesting of wheat crop is 3 to 4 man-h/ha. The unskilled labour requirement for initial field preparation, collection and binding of harvested crop is 40 to 45 man-h/ha. The unskilled labour require-ment in manual harvesting and collection as well as binding of harvested crop works out to 128 man-h/ha and 48 man-h/ha, respectively. Thus, there is a saving of 127-133 man-h/ha with the use of tractor-operated reaper. The field capacity of the machine is 0.4 ha/h when operated at forward speed of 2.5-3.5 km/h.

Fig. 8.5: Tractor front mounted vertical conveyer reaper windrower.

Self-propelled walking type vertical conveyor reaper

The need of a small self-propelled vertical conveyor reaper powered by a light weight diesel engine was felt keeping in view the specific situation prevailing in paddy growing areas. The first prototype of 1.0 m diesel engine self-propelled vertical conveyor reaper windrower operated in India was developed at Coimbatore during 1983 under the CIAE-IRRI Industrial Extension Project financed by Inter-national Rice Research Institute (IRRI) based on the IRRI design of machine powered by a petrol engine (Singh and Pandey, 2008; Pandey *et al.*, 1997). Self-propelled vertical conveyor reaper includes 6 hp light weight diesel engine, crop row

Fig. 8.6: Self-propelled walking behind vertical conveyer reaper windrower in operation

dividers, star wheels, standard cutter bar having 76.2 mm pitch knife sections, vertical conveyor belts, steel lugged wheels and power transmission system (Fig. 8.6). Self-propelled vertical conveyor reaper is used for harvesting cereal crops like wheat, paddy, soybean, saflower, oilseed crops etc. It cuts the crop, conveys it vertically to one side and drops in a windrow for easy collection. The skilled labour requirement for harvesting by reaper is 5 man-h/ha. The unskilled labour requirement for initial field preparation, collection and binding of harvested crop is 40 man-h/ha. The field capacity of the machine is 0.15 - 0.17 ha /h at a forward speed of 2.5-3.5 km/h. The fuel consumption is about one litre. There is a saving of 90-95% in labour, time and cost of operation as compared to conventional method of manual harvesting with sickle.

Self-propelled riding type vertical conveyor reaper

A riding type self-propelled vertical conveyor reaper windrower (Fig. 8.7) has been developed (Pandey *et al*, 1997). The reaper is powered by a 6 hp diesel engine and the effective cutter bar width is 1.53 m. It is provided with four pneumatic wheels, two driving wheels in the front having size of 7.50 – 16, 14 PR and two steering wheels on the rear having size of 4.50 – 10, 8 PR. Simple systems of clutch, brakes, steering and hydraulics have been incorporated and an operator's seat is available to make the machines riding type. It has two forward and one reverse speed. The machine can harvest different varieties of wheat, paddy and soybean. In case of wheat the machine could be operated at an average speed of 3.5 km/h and effective field capacity is 0.26 ha/h with a machine loss of 0.6%. In case of soybean crop, the plant height of the crop harvested could be in the range of 33 to 91 cm. The effective field capacity is 0.344 ha/h and total machine losses 6.76%.

a) Harvesting of wheat by self -propelled riding type reaper windrower

b) Harvesting of soybean crop by self-propelled riding type reaper windrower

Fig. 8.7: Self-propelled riding type reaper windrower

Self propelled reaper binder

Riding type binder reaper has recently been introduced in India by BCS India Pvt Ltd. It can combine the operations of cutting and binding simultaneously. Thus it eliminates the crop collection and binding operation in strenuous stooping posture after reaping by VCR. It is operated by a 4 stroke; air cooled, single cylinder, variable speed 7.5 kW diesel engine (Fig. 8.8). This machine is claimed to be suitable for paddy, wheat, barley and oat (Anonymous, 2010). It has already been tested successfully for mustard crop also. Though the engine has a rated speed of 3000 rpm, but 2000-2300 rpm engine speed is recommended for field operation as this speed range is near to the maximum torque producing zone. The machine has a mechanical constant mesh type gear box with four forward speeds (4.32, 5.84, 7.7 and 10.8 km h^{-1}) and one reverse speed (4.6 km h^{-1}). The average field capacity of the machine for paddy is 0.3 ha/h with the through put of the material 2 ton/h. The height of cut from ground level is 7 cm the saving in cost of operation and labour for harvesting paddy with this machine is 41% and 51% respectively as compare to manual harvesting of paddy.

Fig. 8.8: Self-propelled reaper binder during field operation.
Courtesy: BCS India Pvt. Ltd.

Power tiller-operated vertical conveyor reaper windrower

A power tiller-operated vertical conveyor reaper with 1.2 m size cutter bar has been developed (Fig. 8.9), Anonymous, 2006. The effective field capacity of machine is 0.12 ha/h at a forward speed of 1.64 km/h. The average fuel consumption is 1.1 l/h. Pre-harvesting losses and harves-ting losses are 2.6 % and 6.41 %, respectively.

Fig. 8.9: Power tiller-operated vertical conveyor reaper windrower

Grain combine harvester

A combine is farm machine that combines the reaper and thresher to harvest the standing crop, thresh it and clean the grain from straw in one operation. According to source of power used combines may be classified as self-propelled combines (Fig. 8.10) and tractor operated (Fig. 8.11) or trailed type combine (Singh, 2007; Singh and Verma, 2009; Pandey *et al.*, 1997). Self-propelled combines use a propelling power source to do various operations. In tractor operated combine; the power is being optioned through detachable tractor. Only at the time of harvesting tractors are attached and rest of the times it can be used for other farm operations.

Fig. 8.10: Self-propelled combine working in field (courtesy: CLAAS India Ltd.)

Fig. 8.11: Tractor operated combine working in field

The present day combine harvesters are being mostly used for harvesting two major crops namely wheat and paddy. Other crops can also be harvested with combines like, sunflower, maize, soybean, pulses etc, with slight changes in the combine. A combine harvester consists of header platform, reel, cutter bar, crop divider, platform auger, feeder conveyer, cylinder, concave and grate, fan, chaff sieves, straw walkers, grain sieves, grain auger, tailing auger, grain elevator, grain container and grain unloading auger (Fig. 8.12 and Fig. 8.13).

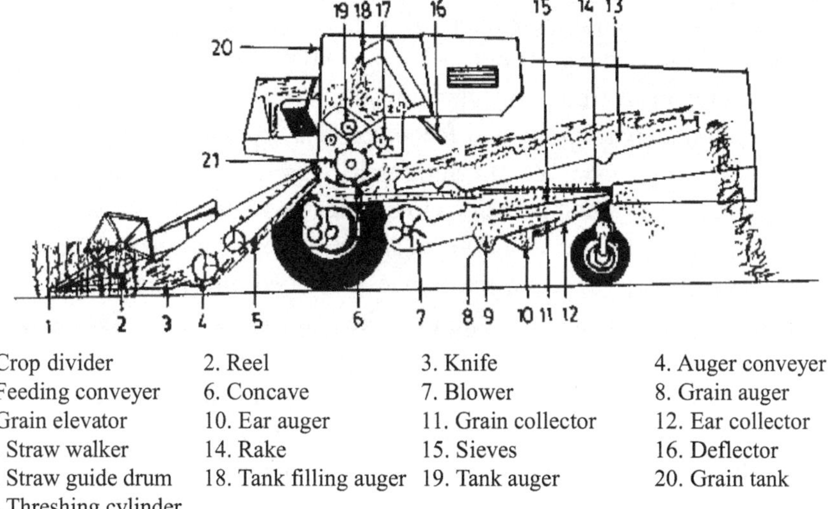

1. Crop divider
2. Reel
3. Knife
4. Auger conveyer
5. Feeding conveyer
6. Concave
7. Blower
8. Grain auger
9. Grain elevator
10. Ear auger
11. Grain collector
12. Ear collector
13. Straw walker
14. Rake
15. Sieves
16. Deflector
17. Straw guide drum
18. Tank filling auger
19. Tank auger
20. Grain tank
21. Threshing cylinder

Fig. 8.12: Details of a self-propelled combine.

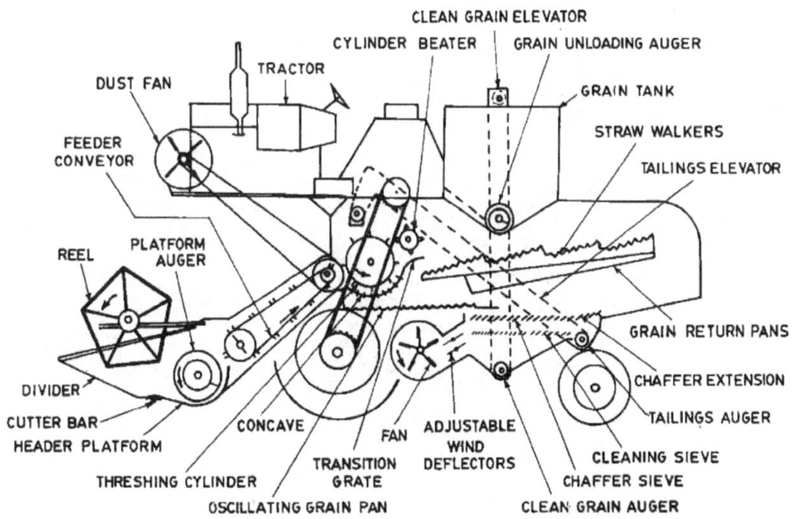

Fig. 8.13: Details of a tractor operated combine harvester.

The function of cutting and feeding unit of combine is to cut only as much of the crop as is necessary to get all the heads, lay the cut crop on the platform or auger and then feed it uniformly into the threshing unit. This unit comprises of cutter bar, reel, and platform auger and feeder conveyor (Fig. 8.14). Threshing unit of combine thresh the grain from the heads. The grains are separated from the pods by impact, rubbing or squeezing actions between cylinder and concave. The threshing cylinder may be of rasp bar type having corrugated bars, angle bar type (right angle bar with rubber facing) or spike tooth type pegs or spikes on it. The concave is a grate composed of rods and bars or wires. It is at the concave grate and finger grate that as much as 90% grain is separated from the crop. Cylinder beater helps in cleaning the straw from the cylinder thus preventing cylinder wrapping and feedback. The separating unit agitates the straw after it comes from the threshing unit. This shakes out the loose grain remaining in the straw and delivers it to the cleaning unit. The straw is carried out of the combine by the rake. This is either one-piece straw rake or walker type with 3 or 5 pieces. Grain return pan is located under the straw rake. It catches the grain as it falls through the rake and moves forward to the grain pan. The grain pan is located under the forward part of straw rake behind and below the cylinder. Its function is to catch the grain from the concave and cylinder grates and form grain return pan or conveyor for delivery to the cleaning unit. The function of cleaning unit is to separate the clean grain and deliver it

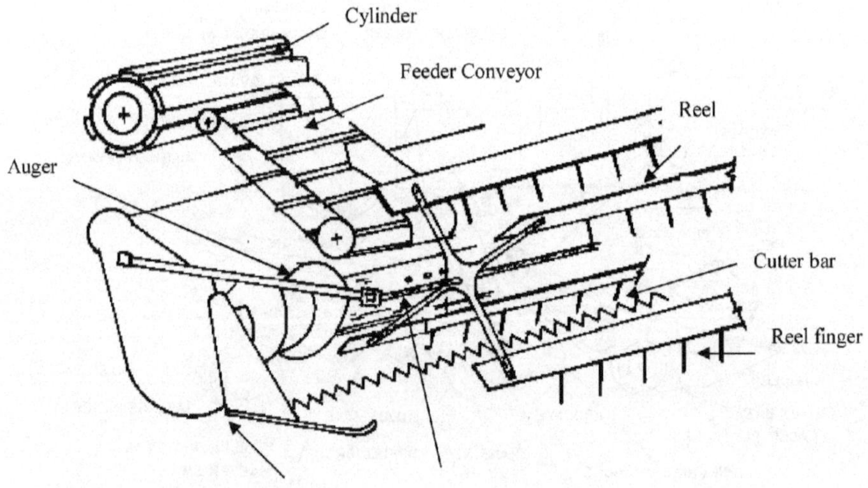

Fig. 8.14: Cutter bar assembly of a combine harvester.

to the grain tank, return tailings to the cylinder for threshing and move the remaining material out of the combine. This is accomplished by means of adjustable chaffer, sieves and cleaning fan bar gravity and air blast. Tailings auger and tailings elevator collects the material which comes off the lower sieve plus any material which falls through the extension chaffer and deliver to cylinder for re-threshing.

The cleaning unit consists of a number of sieves and fan cleans the grain. The unthreshed grains pass through tailing auger and go to cylinder for rethreshing. The grain passes through an elevator and collected in a hopper or directly unloaded to the trailer. The fan is adjusted such that the chaffs are blown off to the rear of the machine. The size of the combine is indicated by the width of cut it covers in the field. Some adjustments are always necessary on combines before being used for harvesting of crop. The height of the forward tip of any knife section above the plane on which the combine is standing expressed in millimeters, is called cutter bar height. This height is adjustable and ranges from 75-900 mm. The crop is cut just low enough to cover nearly all the heads. If straw is to be saved cutting may be at a lower height. It is essential to keep the reel speed in proper relation to the forward speed of the combine. If reel speed is more, grain shattering from heads occurs. If reel speed is lower then grain may fall on the cutter bar. Hence speed of reel should neither be more nor less. The speed of reel can be adjusted by changing the reel-driving sprocket. The tension of upper and lower feeder canvass is kept tight enough to prevent slippage. If they are too tight, there will be wastage of power and excessive wear will take place.

The gap between the tips of the cylinder to inner surface of the concave is called cylinder-concave clearance. The method of adjusting clearance varies with different machines. In some of the combines, raising or lowering the cylinder changes the cylinder-concave clearance, whereas, in other cases, the height of concave assembly is adjusted to change the cylinder-concave clearance. Close clearance results in more breaks up of the straw, which decreases the effectiveness of separating and cleaning unit. The spacing between concave and cylinder at front is usually adjusted from 2-30 mm and at rear 2-18 mm. With close type concave, the normal setting of concave is closer to the cylinder at the front rather than rear. With open type concave, the concave is normally closer at rear than at front. The sieve is oscillated so that grains pass through chaffer openings and chaff and un-threshed materials thrown at rear of the machine. The chaffer opening should be provided such as to float chaff away without blowing out grain, but coarse heavy material should be retained and discharged at the rear. Shoe sieve opening should be just large enough to permit free passage of grain. Adjustable shoe sieve with smaller lips and openings are the most common. Tailing gate is a device provided at the rear of the cleaning unit to prevent un-threshed material from passing out of the combine. It can be adjusted by raising the tailboard at the end of the chaffer extension if necessary to prevent threshed grain from being blown out.

Grain loss: Each of the four separating areas of the combine cutting and feeding, threshing, separating and cleaning can be a source of loss (Fig. 8.15). The losses in these areas are usually known as the cutter bar, cylinder, rack and shoe losses. Cutter bar loss includes heads of grain missed by the cutter bar, grain shattered out of the head as the knife cuts the straw, grain cut and dropped to the ground before reaching the feeding platform, grain shattered out when the reel strikes the standing grain and/or heads of grain thrown out by the reel. These are known as shattering loss. The cylinder can cause loss in two ways i.e. unthreshed grain left in the heads and carried to the rear of the combine by the rack and/or cracked grain in the grain tank caused by the cylinder running too fast or the cylinder concave being too close. The rack loss is the loose grain which has not been separated from the straw as it passes over the rack and is carried out of the machine with the straw. The shoe loss is the grain that is carried over the rear of the sieves with the chaff or blown out of the combine with the air blast by the blower. The four losses taken together will show how good a job of combining you are doing. Pre-harvest loss i.e. grain found on the ground before harvesting is in addition to the above machine losses.

There are different types of grain losses in the field before and during combining of crops. Moisture contents at the time of harvesting affects the grain losses. At low moisture content, grain losses are pre-harvest shattering loss, cutter bar loss and more breakage of grain. At low moisture the straw is broken finely by the cylinder and more material flows to the sieve resulting into separation problem. There is a risk of natural hazards like rain and hailstorm, which also leads to lodging of crop. Due to delay in harvesting, more weed growth takes place that causes choking of combine. At high moisture content, grains are badly damaged by the cylinder action. The threshing is poor and good cleaning is also a problem. This leads to higher cylinder loss and lower cleaning efficiency. The grains get struck to moist straw and are carried away with straw and chaff. There might be choking problem at different stages in the combine due to high moisture content. As per BIS the combine losses should be maximum 2.5% for wheat, paddy and gram and 4.0% for soybean (IS: 8122 Part II – 1981). Various combine losses are discussed below:

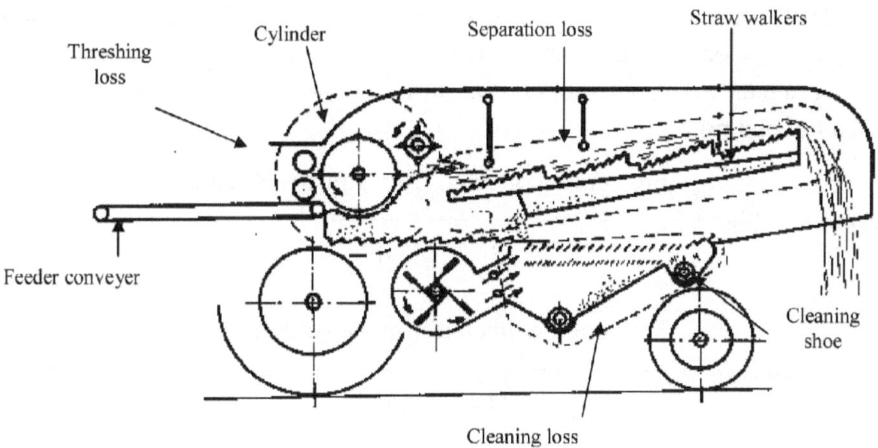

Fig. 8.15: Various combine loss sections on the combine harvester.

Pre-harvest loss: It is determined at minimum of three places randomly selected in the field where combine harvester is to be operated. The sample should be collected from the area having one-meter length in the direction of travel and full or half width of cutter bar of machine depending upon its size. All the loose grains, complete and incomplete ear heads fallen in the marked area have to be picked up manually without vibrating the plants before the machine is to be operated. This will give pre-harvest loss.

Header loss: It is determined on those portions of ground, which are protected from combine afflux by the use of rolls of cloth. The loose grains and complete and incomplete ear heads fallen on the marked area, where pre-harvest losses were determined, shall be collected manually. This gives the header loss. It is also called cutter bar loss.

$$\text{Header Loss (\%)} = \frac{\text{Grain collected from 1 m}^2\text{ area after harvest} - \text{grain collected from same area before harvest}}{\text{Gross yield}} \times 100$$

$$\textbf{Cylinder loss (\%)} = \frac{\text{Unthreshed grain collected from straw rack \& sieve}}{\text{Gross yield}} \times 100$$

Rack and shoe loss: For determining the rack and shoe loss, the straw and chaff afflux is collected separately. To collect these, two rolls of cloth 30 m in length and one and half times the width of straw/chaff outlet is suspended on especially attached fittings beneath the rear of machine. As the sheets of cloth unroll, one sheet retains the afflux from straw walker and other from sieve for 20 m run length. Unrolling operation starts 5 m in advance and terminates 5 m ahead of end point.

$$\textbf{Rack loss } (\%) = \frac{\text{Free grain collected from straw rack sample}}{\text{Gross yield}} \times 100$$

$$\textbf{Sieve loss } (\%) = \frac{\text{Free grain collected from sieve sample}}{\text{Gross yield}} \times 100$$

Grain crackage: It is determined from the samples taken from grain tank. Only visible damaged grains are separated and expressed in percentage of sample taken.

$$\text{Grain crackage } (\%) = \frac{\text{Damaged grain wholly or partially collected from grain tank}}{\text{Gross yield}} \times 100$$

Net yield = Grain collected in the tank/bag from combine test area.

Gross yield = Net yield + cutter bar/header loss + cylinder loss + rack loss + sieve loss

Total combine loss = Cutter bar/header loss + cylinder loss + rack loss + sieve loss

$$\text{Performance efficiency, } \% = \frac{\text{Net yield}}{\text{Gross yield}} \times 100$$

$$\text{Unthreshed } (\%) = \frac{\text{Unthreshed grain in tank + cylinder loss}}{\text{Gross yield}} \times 100$$

$$\text{Cleaning efficiency, } \% = \frac{\text{Clean grain}}{\text{Total grain collected from main outlet}} \times 100$$

$$\text{Threshing efficiency, } \% = \frac{\text{Threshed grain from all outlets}}{\text{Grain output in tank}} \times 100$$

Losses with the best combine adjustment will vary greatly depending upon the type, variety and the condition of the crop. Total losses in clean crops of wheat and barley will vary from approximately 1 to 4% of the total yield. Under good harvesting conditions, the total loss should not be more than two percent. Under average conditions the losses expected from harvesting small grain crops like wheat, paddy, barley etc are 0.5 to 2% cutter bar loss, 0.5% to 1% cylinder loss, 0.2 to 0.4% rack loss and 0.2 to 0.4% shoe loss.

In grain harvesting equipment the forces work during the field operations are shearing force in cutting and straw chopping; impact force in threshing and cutting; rubbing force in threshing and shelling; pneumatic force in cleaning; gravitational force in separation, cleaning and sieving; combination forces - e.g. pneumatic and gravitational in cleaning; frictional and kinetic in conveying, rolling and feeding; centrifugal force in threshing and separation and compression force (squeezing) in threshing. With increase of threshing cylinder diameter from 305 to 533 mm the power requirement decreases substantially. Closed grate requires 25% more power than open grate at cylinder speeds of 23 to 33 m/s. Increasing length of open grate concave from 338 mm to 508 mm increases power requirement by 10%. Maximum engine power rating is 70 to 100 W per mm of cylinder width for self-propelled combine. Total power requirement (including propulsion) for self-propelled combine is 28 to 41 kW at 0.13 to 0.15 kg/min of non grain feed rate per mm cylinder width.

Since combines involve heavy investments, it becomes very important to the users to have the knowledge regarding their proper selection, efficient operation, adjustments, safety precautions, trouble-shooting and maintenance so that they can render efficient and trouble free service to the farmers. Horse power rating, header size and separator size is rough measures of combine size. Though the capacity of the combine is affected by these elements but it cannot be measured solely in terms of dimensions of the combine or its horsepower. Tractor or engine horsepower, type of threshing cylinder, separator width and length (total square centimeter of area), header size and grain tank capacity determines the size and capacity of the combine. The separator width and length govern maximum possible capacity of the combine to separate and clean the crop. Separator width varies from less than 60 cm to over 150 cm and its length varies from about 270 to 430 cm. Threshing cylinder may be either the narrow type with a 50 to 55 cm diameter, or the wide type with a 38 to 40 cm diameter. The small diameter cylinders may have less lugging capacity (ability to continue to rotate at proper speed when heavily loaded) than the larger cylinders. The horsepower of the combine and the size of separator govern header size. Header size ranges from 2.5 to 7 m.

Recent advances in grain harvesting equipment: Change and improvement is a continuous process. Advances have been achieved in the grain harvesting equipment by improving the design of sickles, reapers, vertical conveyor reapers, threshers, multi-crop threshers, and combine harvesters which incorporates all the functions necessary to harvest grains. Some of the recent advances in grain harvesting equipment in terms of their design and functional performance are axial flow combines.

Axial flow combine

In a conventional grain harvesting combine the separation and cleaning units become limiting to increasing its capacity. Grain losses (separation losses) increase rapidly as feed rate increases. The separation process in straw walkers is dependent on gravitational forces necessitating large cross-sectional area and making it slope dependent. These problems have been overcome in axial flow combines where the transverse threshing cylinder rear beater and straw-walkers have been replaced by a longitudinal rotor. Its front portion does the threshing and rear portion the separation. Whereas conventional threshing is single pass process, the axial flow thresher separator is multipass process. The total dwell time of crop in threshing and separation process is less in axial flow combines than in conventional combines. Axial flow combines reduce number of moving parts from 36 to 6 in the thresher separator. Since axial flow concept results in compact design, a larger grain tank can be provided. The centrifugal separator is self-emptying and relatively maintenance free due to high "g" force fields.

Since some grain harvesting equipment performs number of operations simultaneous and continuously, many of these functions cannot be adequately monitored visually or by hearing alone. It is desirable to monitor such functions in a way that abnormal functions can be suitably reported on dashboard. Some of the functions which need monitoring are: cleaning fan, cleaning sieves, tailings, clean grain, straw racks, cylinder rpm, forward speed and grain loss. Automatic control system is desirable for header height adjustment and feed rate as related to forward speed. It is obvious that a number of advances have been made in grain harvesting equipment, which will make it more efficient.

Track type small combine

The track type small combine harvester available in three and four row models are highly suitable for harvesting rice under wet soil conditions as they are provided with rubber tracks for locomotion (Pandey *et al.*, 1997). It consists of crop lifting mechanism with retractable fingers, reciprocating cutter bar of 1 m length, crop conveying system, ear head thresher and cleaning and bagging unit (Fig. 8.16). The crop is held between two chain conveyors, threshed and the remaining straw is released in the field in the form of windrows. It has a provision also to chop the straw and spread it uniformly in the field for subsequent sun drying and incorporation into the soil. It has 4-stroke water cooled diesel. Reaping section lifting is by hydraulic. It can be used for crop length 60-130 cm. Effective field capacity is 0.20 ha/h. It is suitable for combine harvesting rice and other cereal crops.

Fig. 8.16: Different types of track type small combine
Courtesy: M/s. Redlands Ashlyn Motors (P) Ltd., Thrissur

Root crop harvesting equipment

Potato is an important food and vegetable crop, which is also used as an industrial raw material for making starch, alcohol etc. India ranks fourth in area and fifth in production of potato in the world. Increase in potato yield can be attributed to many factors, viz. availability of disease-free seed varieties, irrigation, use of higher and balanced doses of chemical fertilizers, herbicides and pesticides, adoption of improved cultivation practices as well as better power sources and farm equipment. Potato has been grown in India for centuries by traditional methods by using hand tools and animal drawn implements. Modern practices of production demand precision in the placement & utilization of costlier inputs like seed, fertilizer, pesticides etc. Mechanization has played a pivotal role in timeliness of farm operations, reduction of drudgery as well as in cutting down the labour and cost of production for the crop. Like potato, groundnut is another important oil seed crop of India. Digging of groundnut and onion is a critical operation in production. Proper machine for performance of operation and their proper management is essential to get desired results from digging of these crops. As these crops are high value crops, damage to the crops and losses in digging could result in economic loss to the farmers. Timely harvesting of these crops is important factor for getting high yields. Delay in digging may result in left over losses in soil and high cost of digging. Conventionally, traditional tractor operated Duckfoot cultivators or animal drawn desi ploughs were used for digging. In case of manual digging, spade and pick axe are used. Use of these equipment result in damage to the tubers/ bulbs and result in high losses apart from involving high drudgery and cost.

Both tractor operated and animal operated equipment have been developed for digging of these crops (Singh and Garg, 1977; Verma, 1979; Verma and Garg, 1970a, 1970b). Tractor operated equipment include tractor operated digger, tractor operated digger shaker; whereas the animal operated equipment are mainly the diggers. Important components of animal operated digger include beam, main frame with the hitch system for connecting the beam, handle, depth adjustment-cum-ground wheels, digger blade and lifters. Type of blades could be straight or V-shape. Capacity of the unit could vary from 0.05 to 0.12 ha/h. The unit can also be used as secondary tillage equipment. Important components of the tractor operated digger include the main frame with hitch and depth adjustment-cum-ground wheels, shanks, digging blade and lifter rods. The digging blade size could vary from 1.3 to 1.5 m. Field capacity of the unit could vary between 0.35 to 0.4 ha/h. Both animal and tractor operated diggers before use are to be adjusted for width of operation and accordingly the blades are put on the machines and also for depth of operation keeping in view, the likely distribution of the pods/tubers/bulbs in the soil. The machines are properly hitched to the tractor or the animals before operation. Before operation, it should be checked that blades and lifter rods are properly fastened.

Potato harvesting equipment

Manual harvesting of potato requires 600-700 man-h/ha and accounts for nearly one third of the total labour requirement for production of the crop in India. In hand harvesting the ridges are split with the help of a digging tool, viz. spade, hoe, fork or "Ramba" (Khurpa). The tubers are exposed and picked up by hand. Apart from being slow and tedious, manual-harvesting leads to sizeable percentage of cut and bruised tubers.

Potato is grown on ridges or beds. The tubers lie buried under ground while the leaves and stems remain outside. The unit operations involved in potato harvesting are (a) cutting or destroying of the potato haulms before or during harvesting operations, (b) splitting or breaking of soil-potato-ridge and exposing the tubers with a view to collecting them easily and promptly, (c) cutting, lifting and/or conveying of the soil-potato mass with a view to discarding the soil, stems, etc, and collecting the tubers with minimum injuries. Alternatively, soil-potato mass may be scattered with a view to attaining maximum exposure of the tubers thereby enhancing the efficiency of picking of tubers and reducing the labour and cost requirements for the operation. In the latter case, the separation of the potatoes is done by manual picking of the exposed tubers and leaving the soil and clods, roots, etc, in the field itself. The general design requirements for a potato harvesting equipment based on the aforementioned objectives can, therefore, be outlined as:

i) The digger should serve the desired purpose, namely, to split the ridge and expose the tubers or split the ridge and pick and lift the material and separate the soil in a single pass, or split the ridge, pick, lift and convey the soil-potato mass, discard the soil and other extraneous material in the process through manual, mechanical, electronic or any other means or by a combination of these all.

ii) Depending upon the intended purpose, the draft and power requirements of the digger/harvester should be well within the capacity of the source of power to be employed for its operation.

iii) The shape of the digging shovel/share should be such that the draft is low and the flow of the soil-potato mass is smooth.

iv) The digger/harvester should be suitable for a wide range of soil, crop and field conditions.

v) It should be suitable for harvesting the tubers under the prevailing or modified agronomic practices. This is important. For instance, an animal drawn digger can be used without much problems even when the crop is planted at row spacing of 45 cm. However, the minimum row spacing for a tractor-operated digger is 60 cm or more to allow the tractor tyres to move freely in the furrows and not to ride over the ridges.

vi) The working principles and mechanisms of digger/harvester should be such that the potato skinning, cutting and bruising losses during splitting of the ridge and picking, transporting and separation of the material are the minimum.

vii) In a digger requiring manual picking of the tubers from the field, the exposure of the tubers should be maximum so that the time and labour requirement for picking and collecting of the tubers gets minimized.

viii) For a potato digger/harvester, the percentage of unexposed/undug/unlifted tubers should be minimum.

ix) The tubers, in a digger requiring manual picking from the field, should not be scattered over too wide a swath to reduce the picking time and labour requirements.

x) Depending upon power source and the type of the digger, it should dig one or two ridges at a time.

xi) The digger should be suitable for different potato varieties. This is important especially for countries like India where large-tuber varieties are not yet prevalent, unlike the western countries.

xii) It should have necessary provision to adjust and maintain the desired depth of digging to minimize tuber injury and undug tubers.

xiii) The initial cost of the digger/harvester should be affordable i.e. within the purchasing capacity of the farmer.

xiv) The field adjustments should be simple and quick.

xv) The fast-wearing components requiring rebuilding/replacement should be the minimum.

xvi) The design should be simple and it should be possible to manufacture its components out of available materials and with available manufacturing technology and expertise.

xvii) The overall size of the implement should be such that it does not pose any problem in its transportation and turning while in operation.

xviii) The design should be such that it should require minimum attention by way of greasing, oiling etc for its day to day maintenance.

xix) The weight of the digger if it is to be mounted behind a tractor and its rearward projection should not be excessive to be within the permissible hydraulic lift capacity of the tractor.

Haulm cutting equipment

In mechanical harvesting destroying of potato vines/haulm is an essential pre-harvest operation. It offers the advantages of destroying haulm 10 to 12 days before harvesting helping in curing the tubers. In turn, it minimizes the skinning and bruising damage during harvesting. Transmission of virus through leaves and stems, under favourable conditions, is checked and soil-potato separation during harvesting is improved significantly. In advanced countries, where potato cultivation is completely mechanized both chemical and mechanical methods are used for destroying of potato haulm. In the chemical method, aerocynamide dust, ammonium sulphate, aero-sodium cynamide or copper sulphate-sodium chloride is sprayed about two weeks before harvesting to kill the vines. Published information pertaining to Indian experience on chemical vine killing lacking. As regards the mechanical methods, potato haulm pulverisers (cutters) are in common use in overseas countries. They are used for cutting or failing of vines ahead of harvesting. The common practice in India is to cut the wines with the help of sickles, 10-12 days before harvesting. it requires 50-75 man-h/ha. Efforts have been made towards development of haulm-cutting equipment in India (Verma, 1979).

Rotary haulm cutter

It is tractor-mounted equipment consisting of two vertical rotors with discs of 25 cm diameter mounted at the lower ends. Two blades of 36-cm length are rigidly bolted to each disc. Power from tractor PTO is transmitted to the vertical rotors through a gearbox. The blades rotate at a speed of 1500-2000 m/min corresponding to PTO speed of 540 rpm. A fence is provided on the rear and sides to prevent scattering of the cut vines as well as for safety. Gauge wheels are provided to adjust and maintain the desired height of cut. Haulm from four ridges is cut in each pass. The rotor shafts are aligned with the centre of the furrows such that the blades cut the vines from two adjacent ridges. The capacity of rotary haulm cutter is 0.2-0.5 ha/h while operating at forward speeds between 1.5 and 3.5 km/h. Depending upon potato variety and field conditions, the haulm-cutting efficiency varies from 80-90 percent. The main deficiency observed in the rotary haulm cutter is its inability to cope with the vines being on the sloping sides of the ridges. It also cut some tubers projecting over the ridge if the field is not properly levelled. This implement could reduce the labour requirement by 70-80 percent.

Flail-type haulm cutter

This type of haulm-cutter has been in common use in the overseas countries. It employs beaters attached to a horizontal shaft. The length of the beaters is contoured to match the ridge profile. The rotary beaters consist of blades, chain or rubber dampers. The blade rotates at peripheral speeds of 2500-3500 m/min. Prototype of a flail-type haulm cutter using link chains has been developed at PAU, Ludhiana (Verma, 1979). It is mounted on the three-point linkage of a 35-hp tractor and driven by the PTO by means of a gearbox, V-belts and pulleys. The link chains of varying lengths attached to the rotor shaft revolves at a tip speed of 1500-2000 m/min. The height of cut is maintained with the help of the gauge wheels. Haulms from four rows is cut in each pass with an average field capacity of 0.5 ha/h. The uncut vines are found to be under 5 percent. To minimize tuber damage, it is essential to plant the crop with a planter to ensure uniform size of ridges and furrows. Swinging blades of varying lengths can be used instead of the link chains.

Animal-drawn potato digger

It is essentially a potato-raising plough, which splits the ridge and exposes 60-70 percent tubers depending upon the crop and field conditions. It requires 20-25 persons to pick the tubers from an area of one hectare in an 8-hours day. The unexposed tubers are recovered by hand. Picking of both exposed and unexposed tubers is carried out simultaneously or one after the other. Main advantage of an animal-drawn potato digger is its low cost and simple construction, with virtually no adjustments. A traditional potato raising plough throws the soil-potato mass at the sides. Improvements have been made to drop the soil-potato mass at the rear. Gradual curvature of the mould board(s) and provision of upward planting rods at the rear have helped to achieve it.

With this improvement the soil-potato mass is loosened by the share, lifted by the digger body i.e. the mould boards and the extension rods. Potato-soil mass is dropped behind the implement.

It consists of a V-shaped blade having round bars at the rear of blade, tyne, hitch, frame, handle and beam (Fig. 8.17). The round bars are of 12-mm diameter welded at a spacing of 40-mm intervals to form a screen for separation of tubers from the soil (Garg and Singh, 2002). The length of these bars from the cutting point of the blade is kept 40 cm and the ends are bent slightly downward to facilitate the sliding of the soil over the bars. The length of the blade is 54 cm and is made from a one cm thick and 10 cm wide carbon steel flat. The height of the beam can be adjusted from the hitch frame. The implement is operated by an average pair of bullocks and harvests one row completely. The percentage of the tubers bruised, cut or left undug are negligible. The performance of the machine is better when operated at a speed of about 2.0 km/h and at a cutting angle of 31.0 degrees. It can harvest 0.6-1.0 ha/day and a crew of about 20 persons can collect the harvested tubers in comparison to about 50 persons required by "Khurpa". The use of machine saves 40-50% labour and 35-40% in cost of operation over traditional method. It has been reported that an improved animal-drawn digger has 10% less draft and tuber exposure goes up to 70% with the digging efficiency of 98%.

Fig. 8.17: Animal drawn single row potato digger

Tractor operated two row potato diggers

Machine is suitable for harvesting potato tubers and is operated by a tractor. It has 2 blades fitted on the tines (Fig. 8.18). These units are mounted on the cultivator frame having holes for adjustment of rows (Singh and Garg, 1977; Garg and Singh, 2002; Singh and Pandey, 2008). The blades are of 54-cm length and have round bars of 12 mm diameter welded at the back at a spacing of 40-mm interval. The angle of penetration of the blades is adjusted from the top link of the hydraulic system of the tractor. The blades are made from 10 mm thick and 10 cm wide carbon steel flat. The blades are made in such a way that they resemble V-shaped shovels and have an angle of 145 degrees. A 35-hp tractor operates the machine. It harvests 2-rows at a time. After harvesting most of the tubers are exposed but some of them are covered by loose soil. The mounting of the blades on the frame is made in such a way that 2 rows are left undug between the blades. However, at the time of harvesting, a system is adopted so that only one undug ridge remains in between the blades. Digging alternate rows give sufficient time to the labourers for picking harvested tubers and clearing the strips for subsequent run of the digger. It is operated at a speed of about 2.0 km/h and can harvest 2.5-3.0 ha/day. The exposure of the tubers after digging is about 90%. About 10% of the tubers are covered by the loose soil. The use of machine saves 40-50% labour and 30-40% in cost of operation.

Fig. 8.18: Tractor operated two row potato digger

Tractor operated potato digger elevator

Machine is suitable for harvesting and exposing the potato tubers (Garg and Singh, 2002). It comprises of a frame, shovel type-digging blade of 55-cm width, endless rod chain conveyor, gearbox, two gauge wheels, idlers and driving sprockets (Fig. 8.19). The elevator conveyor is made of MS rods of 13-mm diameter, which are riveted/bolted to two endless flat belts. The pitch of the conveyor links is 25 mm. The length of the conveyor is 150 cm and it makes an angle of 20 degree with the horizontal. The digger blade harvests one row at a time. Soil-potato mass is picked up and lifted by the chain conveyor. Two agitator sprockets oscillate the conveyor chain rod, which helps to separate the soil. Potato tubers with no or very little soil/clods are dropped on the ground behind the digger. Thus, the tubers are completely exposed which helps in speedy manual picking. The digger gives the best performance when it is operated at forward speeds between 1.5-2.5 km/h and with conveyor speeds

15-20 percent higher than the forward speed of the digger. The effective field capacity of the digger varies from 1.3 to 2.0 ha/day. The tuber damage under normal conditions seldom exceeds one percent including bruised liners. Exposed tubers vary from 95 to 99 percent. About 1 to 5 percent tubers remain covered by the loose soil. Total labour requirement for digging and collecting the tubers is about 150 man-h/ha as against about 600 man-h/ha for manual digging. It saves 70-75% labour and 45-50% in cost of operation. It increases the yield by 4-5% by reducing the losses.

Fig. 8.19: Tractor operated potato elevator digger
(*Courtesy*: M/s. Droli Mechanical Works, Moga)

Depending upon the soil type, digging shovels of different shapes are used in the elevator type diggers. The single-row digger is recommended for a tractor of 25-35 hp and a two-row digger for a tractor of 40 or higher horsepower. Main advantage of an elevator type digger over other types of diggers is the low tuber damage. The skinning and bruising damage under normal conditions does not exceed 2 percent. Furthermore, this type of digger has lower labour requirements due to high percent of tuber exposure (above 95%) as well as the dug tubers being dropped over a narrow band. It is worth mentioning here that the annual repair and replacement charges for the operation of an elevator type digger are fairly high due to rapid wear and tear of the conveyor links, driving and agitator sprockets as well as the conveyor. Use of better quality material for the fast wearing components can help improve the durability and reliability of the elevator type diggers and reduce the cost of operation. Elevator type digger is provided with gauge wheels to adjust and maintain the depth of penetration of the digging blade. The depth is so adjusted that the blade cuts the soil 25-30 mm below the tuber-zone. The depth can be adjusted with the help of tractor top link and controlled by the gauge wheels. It has been reported that every 2.5 cm of extra depth of cut leads to lifting of over one ton of extra soil per hectare. This reduces the separation efficiency and increases the power requirements of the digger. Ideally speaking the digging shovel should run just under the tuber zone. Table 8.1 gives the comparative performance of different types of potato harvesting equipment. The total labour requirements of an elevator-type digger range from 139 to 145 man-h/ha as compared to over 300 man-h/ha for the spinner-type digger and over 400 man-h/ha for the animal and tractor-drawn digging blades (Verma, 1979; Verma *et al.*, 1977).

Oscillating potato diggers

Oscillating or vibratory diggers are among the new developments in the potato harvesting machinery. It has been reported that draft reduction of the order of 40-50 percent can be achieved with an oscillating digger (Verma *et al.*, 1977). The soil-separation efficiency is also significantly improved as compared to a fixed-blade digger. It has been found that an oscillating potato digger works best when the frequency of oscillations is between 8 and 10 cps and the amplitude of oscillations leads to better soil breakup and soil-potato separation but balancing of inertial forces due to rotating and reciprocating masses has been the major problem encountered. Depending upon the forward speed, frequency, amplitude and direction of oscillations, 95-98% tubers can be exposed with the working of an oscillating digger. Relatively simple configuration, light weight, low-cost and compactness are among the main attributes of oscillating digger. However, such a digger requires careful design and balancing of the inertia forces to prevent transmission of vibrations from the digger blade to the digger frame and tractor links. Field capacity of a single-row-oscillating digger has been found to be compatible with the single-row elevator digger with a passive blade. Tuber injuries are, however, much less.

Potato harvesting combine

A potato harvesting combine dismantles the potato ridge, picks up the soil-potato mass, shifts out the loose soil and places the material on a slow-moving sorting conveyor to manually separate the unseparated clods and other foreign matter. Essentially, a potato combine comprises a digging shovel with one or two pieces, or two inclined concave discs to dismantle the ridge and a primary conveyor to pick and elevate the soil-potato mass. Bulk of the loose soil is shifted out by the conveyor with the upper web supported over the agitator sprockets at one or two places. The material conveyed by the primary conveyor is fed on to a cross conveyor or a rotary cage-screen before delivering it onto a sorting conveyor belt. The purpose of the intermediate conveying system is to

Table 8.1: Comparative performance of different methods of potato harvesting.

S. No.	Particulars	Hand digging	Bullock-drawn digger	Tractor-drawn digging blade	Tractor-drawn potato spinner digger	Tractor mounted one-row potato elevator digger	Tractor mounted potato two-row elevator digger
1	Output ha/day	0.01	0.8	2.5	1.6	1.4	2.8
2	Digging						
	Man-h/ha	650	6	4	5	5	3
	Bullock-h/ha	-	20	-	-	-	-
	Tractor-h/ha	-	-	4	5	6	3
3	Picking exposed tubersMan-h/ha	-	360	360	288	80	80
4	Picking tubers in racksMan-h/ha	40	40	40	40	40	40
5	Planting of covered tubersMan-h/ha	-	16	16	16	16	16
6	Total labour requirementMan-h/ha	690	422	420	349	142	139
7	Bruising of tubers (%)	2.0	1.5	1.5	3.0	1.5	1.5
8	Cutting of tuber (%)	5.0	0.5	0.5	1.5	1.5	1.5

effect further separation of the soil and feed the material onto a slow moving conveyor to enable manual picking of the remaining foreign matter. In some designs, the sorting belts have provision to alter the slope with the horizontal to separate the tubers from the clods and other foreign matter by using their differential repose angles. Potato tubers have smaller angles of repose than the clods. Electronic sensors have been used in potato combines for separation of tubers from the soil. Such machines are, however, quite expensive. Potato combines used in overseas countries are of single or two-row types. A potato combine works best on potato varieties with large tubers planted in relatively bigger fields with proper field layout. For efficient operation of potato harvesting equipment the following points should be adhere too;

- Plant the crop in straight rows at 60 cm or wider spacing by means of a potato planter.
- Plant the crop lengthwise to minimize the number of turning at the ends.
- Use a ridger for second earthing up if necessary.
- Provide adequate headland for turning of the tractor.
- Destroy the vines (haulms) about 2 weeks prior to harvesting.
- Add sufficient soil amending substances to develop a proper texture in the soil of the potato fields to eliminate/minimize clod formation.
- Ensure adequate moisture in the soil at the time of harvest to avoid clod formation.
- Use right type of agitator sprockets to effect maximum soil separation without causing excessive injuries to the tubers.
- Ensure that the elevator conveyor web is neither too tight nor too loose. Links of a loose conveyor override the teeth of the driving-sprockets and that of extra tight conveyor cause undue wear of conveyor links.
- Maintain correct tension to minimize belt/chain slippage in the power transmission system.
- Lubricate the shaft bearings every week and those of the telescoping drive shaft daily.
- Check oil level in the gearbox daily before the start of the operation and top up when necessary.

- Alight the digger with the ridger(s) to keep the digging blade in the centre of the ridge(s).
- Avoid excessive depth of digging. The shovel should run 20-25 mm below the tuber zone. Adjust the depth with the help of the tractor top link and the gauge wheels. Digger frame should remain horizontal when the blade is operating at the desired depth.
- Lift the digger at the turns.
- Use correct forward speed. Never run the digger too fast otherwise per-cent exposed tubers will be reduced and the labour requirements will increase.
- The conveyor speed index (CSI) should be kept between 1.2 to 1.3. In other words, the speed of the elevator conveyor should be 20-25 percent higher than the forward travel of the digger.
- Keep critical and fast-wearing components like conveyor links, agitator sprockets, and nuts and bolts handy to be used in the hour of need.
- Harvest the crop in alternate rows if adequate labour for picking the tu-bers is not available.
- Store the digger under a shed during slack season. Apply mobil oil/grease on the blade edge and other parts to avoid corrosion.

Tractor mounted onion harvester-cum-elevator

Onion harvester-cum-elevator has been developed for digging onion and other root crops (Anonymous, 2010, 2012, Singh and Pandey, 2008). It consists of a digger blade made from high carbon wear resistant steel (Fig. 8.20). The width and thickness of the blade is 1144 mm and 16 mm. The blade is mounted on the machine at an angle of 20° with the horizontal. An elevator chain conveyor has been attached behind the blade. The spacing between the MS rods used for the fabrication of the elevator conveyor is 20 mm. The slope of the elevator conveyor is kept at 18°. Two oval agitators are provided in the conveying system for separation of soil particles from the onion bulbs. The power to the elevator conveyor has been provided through a gear box (speed ratio 5:27). Two coulter discs are provided in front of the blade at the outer ends, which helps in easy slicing and lifting of soil by the blade. The field capacity of the machine is 0.28, 0.24, 0.21 and 0.21 ha/h for digging carrot, potato, garlic and onion crop respectively when operated at a speed of 2.78, 2.41, 2.10 and 2.10 km/h whereas

Fig. 8.20: Onion harvester being used on different crops.

respective damage is 1.98, 1.92 1.22 and less than 1.0 percent respectively. The saving in labour ranges from 62 to 71 percent. The performance of the machine is highly satisfactory for digging these crops except excessive soil coming over the conveyor. Saving in cost of operation and labour for harvesting onion, carrot and, garlic is 52.28, 46.71, 52.28 percent and 69.05, 59.29 and 69.05 percent respectively as compared to manual harvesting. There is no saving in labour and cost in case of potato digging as compared to digging of potato using potato digger. However due to more use per year the overall cost of operation will reduce.

Expanding pitch rubber spool potato sizer

It is suitable for grading potato tubers into different grades (Verma, 1979). It comprises of a frame, an elevator feed conveyor, an intermediate receiving conveyor with rubber spools, two identical driving rollers with helical grooves of gradually increasing pitch, collecting platform with adjusting partitions and gates, fenders, transport wheels and a power transmission system (Fig. 8.21). The elevator conveyor is inclined at an angle of 38 degree with the horizontal and is made of 13 mm rods with a pitch of 25 mm. The rubber spools of the sizing conveyor are mounted on 13-mm mild steel rods and two endless link chains at a spacing of 102 mm. The rods carrying the rubber spools rest on the mild steel railings on both sides. Power to the sizer is provided by 2-hp electric motor mounted on a platform. The sizer can grade the potatoes into four different grades. The performance of sizer is better at a feed elevator conveyor speed of 31 m/min, intermediate conveyor speed of 61.2 m/min and helical driving shafts speed of 7.5 m/min. Total damage by way of bruising, skinning and cutting is less than 1 percent. The capacity of the sizer is 2.5 t/h. The performance of the grader in terms of sizing accuracy is observed to be better for potato tubers between 20 and 60 mm diameter. Coefficient of variation is the maximum for the fourth grade which comprise tubers of larger size (7.5 cm and above).

Fig. 8.21: Expanding pitch rubber spool potato sizer

Groundnut harvesting equipment

Groundnut is a root crop and an important oil seed crop. It is grown during Kharif season. However summer groundnut variety has also been developed. It is grown in light to medium to heavy soil. Main groundnut growing states in India are Gujarat, Tamil Nadu, Andhra Pradesh, Madhya Pradesh, and Maharashtra. The crop is planted on flat land as well as ridges & beds. Row spacing is kept 30 cm with a plant to plant spacing of 20 cm. The main requirements for groundnut harvesting are: i) to cut the tap roots slightly below the pod zone, ii) to loosen the soil up to the pod zone such that it is easy to pull out the vines (plants) along with the pods manually or lift the vines by an elevator shaker conveyor and place them in a windrow such that the maximum number of pods get exposed to the sun, for rapid sun drying, iii) to minimize the pod loss, iv) to shake-off the soil and v) to lay the crop in a windrow for subsequent combining.

In the traditional method of groundnut harvesting, the vines are uprooted with a hand tool along with the entire root system. Groundnut is a legume and tap-rooted crop. The groundnut pods are located up to a depth of 5-10 cm usually referred to as pod zone. The manual harvesting leads to avoidable depletion of soil fertility due to removal of the complete root-system bearing the nitrogenous nodules. It requires from 120-150 man-h/ha to harvest the crop. Mechanical harvesting of groundnut has the advantage of reducing the cost and labour requirements and is conductive to better soil fertility as the blade of the digging machine cuts the roots below the pod zone and leaves the remaining root system in the soil itself (Verma and Garg, 1970a, 1970b, Singh, 2007). Harvesting by machines, however, require that the fields should have sufficient moisture to enable the blade to penetrate to the desired depth. Also it requires the crop to be harvested without letting it over mature. Common type of groundnut harvesting equipment is full or half-sweep type blades, sweep-type blade with a pressing roller, digger shaker windrower and groundnut combine. In all these equipment, a single-piece blade slightly curved in the centre is used. Further, disc coulters are provided on both sides of the blade to make a pre-cut on the crop being loosened by the blade. In the absence of coulters, the vines extending outside the blade width are dragged which leads to detachment of pods from the vines. The action of a groundnut-digging blade also results in thorough cultivation of the soil.

Tractor mounted digging blade

The digging blade is mounted on an existing cultivator frame. It works on the principle of a sweep. The depth of working is adjusted with the help of the gauge wheels and tractor hydraulic lift. The vines are cut below the pod-zone and are left in their places without disturbing them. The vines are later collected manually and stacked in bunches in the field for 3-4 days for sun drying before transporting them to the threshing floor. The tractor mounted digging blade is suitable for harvesting groundnut from light soil where adhering of the soil to the pods and plant roots is not a problem. In heavier and harder soils, it does not work satisfactorily due to clod formation for which there is no provision to shake off the soil.

Tractor-operated digging blade with corrugated roller

A hollow corrugated roller measuring about 40 cm in diameter, made of angle-iron sections is attached behind the blade. The length of the blade is kept about 4 cm per hp. The depth of working is adjusted with the help of the top link of the tractor and by adjusting the corrugated roller up or down. The roller in this way serves the purpose of the gauge wheel and helps in removal of the soil adhered to the roots and also minimizes chances of clogging as the vines get pressed after digging. It is not suitable for wet soil conditions as the soil has a tendency to adhere to the vines while being pressed under the roller. As in the case of the digging blade, gathering of the vines into small bunches is carried out manually. This equipment in recommended for digging the crop from relatively heavier soils.

Tractor mounted groundnut digger elevator

Tractor operated groundnut digger consists of a single blade mounted on frame. It moves beneath the pods in soil and after its operation vines could be easily pulled by hand and inverted manually for sun-drying in the field. Labour requirement for this operation is approximately 200 man-h/ha. The time of digging groundnut coincides with harvesting of paddy and other crops. Therefore, sometimes non-availability of labour causes unwanted delay and subjects the crops to losses due to birds, pests and hazards. A tractor operated groundnut digger-cum-elevator has been developed (Garg and Singh, 2002; Anonymous, 2010, 2012). It comprises of a blade, elevator-cum-pickup reel, fenders, gauge wheel, coulters and power transmission system (Fig. 8.22). The front end of the pickup cum elevator reel is adjustable in accordance with the depth of working of blade. The frame, made up of angle iron, has provision of mounting blade, coulters, gear box and three point linkage in the front. Its rear portion is inclined at an angle and pick up cum elevator assembly is mounted on it, which consisted of 15 cross members bolted on the endless chains running on sprockets. The power is transmitted to the rear drive shaft of machine from tractor PTO, which in turn rotates the chain driving sprockets. Two MS shanks one on each side, are mounted on the frame. A single piece blade made up of high carbon steel is bolted with the lower end of these shanks. The working depth of the machine is adjusted with the help of two gauge wheels and the top link of the tractor. The front end of the pickup rod is so adjusted that the spikes comb about 3 cm of the top soil to lift vines gently from the loosened soil. The groundnut vines are cut below the pod zone and simultaneously lifted by a shaker-conveyor. The soil attached to the vines is shaken off in the process and the crop is dropped at the rear in the form of a narrow fluffy windrow. The vines are dropped in such a manner that the pods get exposed to the sun for speedy drying. It requires a tractor operator and one attendant for its working. The machine can be operated with tractors 35 hp and above. The digger cum elevator has field capacity of 0.46 to 0.47 ha/h with the digging efficiency of 97 to 98.5 per cent. The machine saves 65% labour and 32% cost of operation.

Fig. 8.22: Tractor operated groundnut digger elevator

ANGRAU Hyderabad has also developed tractor operated groundnut digger shaker (Fig. 8.23) that consists of digging blade and crop collecting unit (Anonymous, 2010). The length and width of the blade are 1350 cm and 0.8 cm respectively. It is made of M.S plate. The weight of the blade guntaka is 122 kg and is operated by a 35 hp tractor at a speed of 2-3 km/h. The trough type crop collecting mechanism is made of M.S. sheet of 0.5 cm thickness with a length of 90 cm. The front and back width are 140 cm and 45 cm respectively. It can cover four rows at a time. The digging capacity of the machine is 0.15 ha/h with digging efficiency of 90-95%. The comparative performance of different groundnut harvesting machines is given in Table 8.2.

Fig. 8.23: Tractor operated groundnut digger shaker.

Table 8.2: Comparative performance of different type of groundnut harvesting equipment

Particulars	Digging blade	Digging blade with roller	Digger-shaker windrower
Size of blade (m)	1.2-1.5	1.2-1.8	1.2
Tractor hp	20 or more	30 or more	30 or more
Principles of working	Digging the vines and leaving them in their place	Digging the vines pressing them with a corrugated roller to remove the soil from them	Digging, lifting, shaking and windrowing of the vines simultaneously.
Output (ha/day)	2.5-3.0	2.5-3.0	2.5-3.0
Average pod detachment (%)	3-5	3-5	3-5
Labour requirement (Man-h/ha)	30-40	25-30	6-8

Tractor mounted turmeric digger

Turmeric is one of the major spices cultivated over a large area. It is grown largely in Andhra Pradesh, Maharashtra, West Bengal, Tamil Nadu, Kerala and Assam. The general practice of harvesting is to dig out the rhizome manually with the help of hand tools. This type of harvesting causes damage to rhizomes. It is also difficult task on the part of the farmers to get the required labour force during the harvesting season for timely harvesting of the crop. Traditional method employed for digging out the turmeric rhizome is time consuming and injury to the rhizome is unavoidable. To overcome these problems, a machine has been developed. The turmeric digger (Fig. 8.24) consists of a main frame made of MS channel sections and three point link (Anonymous, 2008). Digging blades are of sweep type and can be mounted on the main frame. A front ridger is also provided for digging cross bunds for smoother operation.

Fig. 8.24: Tractor mounted turmeric digger

Tractor operated cassava harvester

Harvesting of cassava crop is very laborious and time consuming. To eliminate this, a tractor operated cassava harvester has been developed

(Anonymous, 2013). It consists of main frame, shanks, digging blade, hitching frame and depth adjustment wheels (Fig. 8.25). It can cover one row at a time but it can be designed for digging two rows also depending upon power availability. The shank has a bent leg plough with an angle of 150 degrees to accommodate the dug cassava tubers. The blade angle of 5 degrees is provided for easy penetration into the soil. The row spacing can be adjusted by moving the shanks on the main frame. The field capacity is 0.08 ha/h for single row. The damage to the tuber is less than 1% and undug tuber 2.5%.

Fig. 8.25: Tractor operated cassava harvester

Maize harvesting and shelling equipment

Maize is an important crop grown for grain, fodder and silage. In India it was introduced in the seventeenth century and the seed was brought from USA. Harvesting and shelling methods depend upon the final use of the crop. In case it is used as fodder/silage, it is harvested at the time of appearance of silky hairs and plant-stalks are green. In case the crop is used as grain crop, it is allowed to ripen till the cobs become yellowish and hard. Then these are cut from the plant stalks and dried in the sun. The grains are threshed from the cobs by beating them with sticks and cleaned and grain manually removed. This traditional method is followed by a majority of farmers throughout India. However medium and large category farmers have started using modern innovations and methods.

Maize shellers in India were introduced in the early 60's. Both manually operated as well as power operated shellers were introduced (Singh, 2010). Power operated shellers have become popular because of their high shelling capacity, low cost and machine simplicity. Power operated sheller however, requires maize crop to be dehusked. Dehusking itself is very labour intensive operation. For shelling of maize cobs, manual (octagonal tubular maize sheller, hand maize sheller, rotary maize sheller) and power operated maize shellers have been developed and commercialized. Maize being a major crop grown in both the agricultural seasons, i.e., *Kharif* (June-September) and Rabi (November to March), in Rajasthan state of the country, work on farm equipment for dehusking, shelling and dehusking-shelling started at many institutions since 1970 (Singh, 2010). Mudgal *et al.* (1998) have reported the development of equipment related to dehusking and dehusking-shelling such as hand operated maize dehusker, pedal operated maize dehusker, pedal operated maize dehusker-cum-sheller, power operated maize dehusker and power operated maize dehusker-sheller.

For making dehusker, a pair of rubber and steel rollers has been used with opposite rotation of each roller. MS rod has also been tried by welding spirally for this purpose on steel roller. Some has used serrated blades lengthwise to facilitate the dehusking. For making dehusker-cum-sheller, wooden cylinder with rubber strips along its length on periphery has been used for shelling. A combined unit for dehusking-shelling in one cylinder has been tried by using half of the cylinder length with rasp bars and the other with rubber strips in octagonal cylinder to act as dehusker and sheller, respectively. In power operated maize dehusker-sheller, helical louverers have been provided for easy passage of dehusked cob after shelling from cylinder. PAU Ludhiana and TNAU, Coimbatore has also worked on maize dehusker-sheller for removing the outer shell and for shelling the maize cob simultaneously.

Manual equipment

Sickles

Harvesting of maize is still done by a majority of farmers with traditional type sickles. However improved versions of sickles have come in the market due to developmental research. The introduction of serrated type blades (hack saw type blades) have improved the harvesting efficiency because the blades offer less cutting resistance thereby contribute to lesser manual effort. This results into greater output. The maintenance requirement in terms of teeth cutting is also low as compared to earlier/traditional sickles. Sickles with hacksaw type blades and ergonomically designed handles are becoming popular and have gained wide acceptability among the farmers.

Manual maize shellers

Metal maize shellers are available in two types of models i.e. RAU Model and CIAE-Model designed and developed at Rajendra Agricultural University, Pusa (Bihar) and Central Institute of Agricultural Engineering, Bhopal (Pandey *et al*., 1997). The RAU model of hand operated maize sheller is made of 20 gauge G.I. sheet of 20 cm. x 10 cm size. The sheet is folded into a cylindrical shape of 6-cm diameter and 14 cm length. The joining ends are fixed together with the help of two rivets. Four rows of 'V' shaped notches with four notches in each row are cut at both ends of pipe. All the notches are inverted inside the pipe at an angle of 90 degree. The lengths of notches are respectively 12.5, 15.0, 17.0 and 19.0 mm for the first, second, third and fourth row respectively. The differences in the length of the notches make the device suitable for various sizes of maize cobs. One end of the sheller is suitable for bigger diameter cobs and the other end for small diameter cobs. The sharp edges of the pipe ends are folded for safety. After the insertion of the cob, the sheller is rotated in one direction and the cob in the opposite direction alternately and thus the kernels are removed. Light force is required to insert the cob inside the maize sheller to avoid grain damage. The maize sheller gives an output of 15 kg of shelled grain per hour and it is 2½ times more efficient than hand shelling.

Fig. 8.26: Manual maize sheller

CIAE model of hand operated metal maize sheller is made from a short length of 62.5 mm dia MS pipe of 16 SWG (Fig. 8.26). Four fins are made from an 18-gauge MS sheet and bent at 70°. These fins are tapered and fitted inside the tube. The tapered ends provide two variable openings - 39 mm at one side and 26.5 mm to the other side. Different sizes of cobs can be handled by this arrangement. By holding sheller in the left hand and inserting the cob with right hand and then rotating in opposite directions, the shelling is done with the kernels removed by the fins. The shelling capacity is around 15 kg of kernels per hour.

Hand operated maize sheller

The maize sheller is operated manually by a driving handle, which rotates the flywheel (Singh, 2010). The flywheel in turn rotates the driven pulley on the shaft on which is mounted the axial flow cylinder (Fig. 8.27). The threshing mechanism separates the grain from stalk or cob through revolving cylinder and concave. The grain is rubbed against drum and is shelled as the cob moves to thresher end. Adjustments are provided for varying concave clearance. Separate outlets for grain and cobs are provided. The capacity varies from 30 to 40 kg of kernels per hour. The domestic sheller is a small hand operated tool suitable for shelling small quantities of maize. It can easily be mounted on a box. The feed opening is provided with spring to take different sizes of cobs. The picker gear, after shelling, directs the cobs to the outlet.

Fig. 8.27: Hand operated maize sheller

Hand operated maize dehusker-sheller

Maize is grown in all the seasons, i.e., *Kharif*, *Rabi* and summer. Of these three seasons, nearly 90 per cent of the production is from *Kharif* season, 7-8 per cent during *Rabi* season and remaining 1-2 per cent during summer season. This crop is emerging as most important crop after rice and wheat. Maize crop sown for grain is harvested when it reaches physiological maturity and contains 25-30 per cent moisture. To provide options before small and hill farmers, a hand operated maize dehusker-sheller has been developed (Singh, 2010). The axial-flow maize dehusker-sheller consists of threshing cylinder (4 beaters with 3 solid lugs on each beater as threshing elements), louvers, main shaft, bottom sieve with equi-spaced square MS bars, top and bottom cylinder cover,

Fig. 8.28: Hand operated maize dehusker-sheller

hopper, trapezium shaped grain outlet, husk outlet, trapezium shaped frame, chain-sprocket system for power transmission etc (Fig. 8.28). The concave clearance is 35 mm. The diameter of cylinder is 380 mm. The average dehusking efficiency is 100 per cent at feed rate of 80 kg/h. The lowest grain breakage is 0.45 per cent at cylinder speed of 5.58 m/s for the feed rate of 120 kg/h. The optimum cylinder speed is 5 to 5.6 m/s while operating the maize dehusker-sheller. The dehusking efficiency, shelling efficiency and grain breakage are 99, 98 and less than 1 per cent respectively at the feed rates of 80 to 100 kg un-dehusked maize cobs/h and 14.53 per cent m.c (d.b).

Dehusking efficiency: Un-dehusked maize cobs received from outlet end or in cylinder after termination of each trial is weighed and dehusking efficiency is calculated using formula

$$\text{Dehusking efficiency} = \frac{\text{Weight of un-dehusked cob received from outlet of machine, kg}}{\text{Total un-dehusked cob fed, kg}} \times 100$$

Shelling efficiency: The grain remained in the dehusked cobs received from outlet end or in cylinder after termination of each trial is weighed and shelling efficiency is calculated for cob-wise and grain-wise using formula

$$\text{Shelling efficiency}_{(\text{cob})} = \frac{\text{Weight of dehusked cob having grain from outlet of machine, kg}}{\text{Total un-dehusked cob fed, kg}} \times 100$$

$$\text{Shelling efficiency}_{(\text{grain}} = \frac{\text{Weight of grain from shelled dehusked cob, kg}}{\text{Total grain received from fed un-dehusked cob, kg}} \times 100$$

Throughput capacity: The weight of grain is obtained during dehusking-shelling of un-dehusked maize cobs in total time of run of machine and throughput capacity of machine is calculated using formula

$$\text{Throughput capacity, kg/h} = \frac{\text{Weight of grain obtained, kg}}{\text{Total time of run of prototype, min}} \times 60$$

Power operated maize shellers

The functionary principles are similar to those of hand operated maize sheller. The additional adjustments are provided for varying concave clearance and cylinder speed (Garg and Singh, 2002). A maize sheller, who requires 3 to 5 hp, usually has drum diameter of 30-36 cm, drum length of 80-100 cm with a cylinder speed of 500 to 600 rpm (Fig. 8.29). There are two types of maize shellers i.e. Spring type and Cylinder type. In the spring type sheller, the cobs are fed into an opening leading to a rotating fluted cylinder, a rotating disc and a spring pressure plate. The kernels are removed from the cobs as they move in between the cylinder and the disc. It may have a blower situated at the bottom to deliver clean grain. There are various sizes of this type available in the market e.g. single hole sheller and double hole sheller. The single hole sheller is generally provided with a ratchet type hand crank to avoid the danger of injury to the operator. A fan is provided to clean the grain from dirt and foreign material. Power operated maize shellers of this type have capacity of 220 kg/h. The double hole sheller is similar in construction as the single hole sheller except that rotating disc in this case has teeth on both sides. The cobs being fed from the two holes are shelled simultaneously. The feed openings are provided with spring plates to take different sizes of cobs. The double hole sheller when operated through a 1.5 to 2.0 hp motor gives an output of 400-450 kg/h. Cylinder type sheller consists of a cylinder with lugs, concave assembly and a blower unit. The lugs on the cylinder are placed spirally to advance the cob through shelling mechanism. The lugs are assisted by two spiral ribs to accelerate the movement. The cobs are fed between the concave and the cylinder and kernels are removed through the action of lugs. The blower with a strong blast of air removes small pieces of cobs and other material from the kernels. Cylinder type shellers are available with different sizes operated by 5 to 10 hp with output range of 8 to 18 q/h.

Fig. 8.29: Power operated maize sheller

Power operated maize dehusker-cum-sheller

The PAU, Ludhiana has developed two types of maize dehusker-cum-threshers namely spike tooth type (modified version of wheat thresher) and axial flow type (modified version of sunflower thresher) for threshing maize along with the husk (Garg and Singh, 2002). The axial flow sunflower thresher that has been modified for maize dehusking-cum-shelling includes the modifications as below:

i) An arrangement to vary the opening of feeding chute to regulate the feeding rate is provided as per crop moisture content.

ii) Peg type of cylinder instead of flat bar and 5 cm cylinder concave clearance. The pegs on the cylinder are mounted in helical pattern such as to form an auger for the movement of crop 12.7 cm/turn.

iii) Concave opening is changed from 6.35 x 6.35 cm to 5.1 x 5.1 cm to suit maize shelling.

iv) The cleaning sieves are blinded from the sides to avoid the side spillage.

v) An upper concave is provided in the first 1/3 length of the threshing cylinder as a de-husking zone.

The machine has threshing cylinder with peg type drum. It uses centrifugal type blower. There are three round sieves with 12.5 mm size hole in upper sieve, 7 mm in middle hole and 3 mm in lower sieve. Concave clearance is kept about 50 mm and opening size of concave is 5.1 x 5.1 mm. Cylinder speed is kept at 13.7 m/s. The output of machine varies from 16.9 to 20.5 q/h at moisture content of 11.8 to 24.8% with an average material capacity of 18 q/h (Fig. 8.30). The output capacity decreases with increase in moisture content. The cleaning and threshing efficiency varies in the range of 94.1-98.7% and 96.0-98.0%, respectively. The overall grain losses vary from 0.65 to 1.42%. The percent recovery of grain with machine is 97 percent. There is a saving of about 60-70% in cost and about 90% saving of labour as compared to traditional method of dehusking and threshing.

a) Tractor operated b) Electric motor operated

Fig. 8.30: Power operated maize dehusker-cum-sheller in operation.

Whole plant maize thresher

Maize is a very predominant crop of southern Rajasthan. This is very popular crop and grown by farmers of all category of land holdings. In some districts it is also grown in rabi season though it is basically a kharif crop. Maize is also a crop for establishment of agro industries because many agro industrial products can be prepared from maize. The harvesting and threshing of maize is time and labour intensive operation spread from Sept to November every year. Since the land holding pattern in this region is mostly from small to medium category of farmers and mechanization is of mixed pattern type mainly animal operated plus tractor operated system. The regular occurrence draught in the region creates shortage of fodder requirement for cattles. During scarcity the fodders are supplied by neighbouring states like Punjab & Haryana through trains & trucks. In some areas of southern Rajasthan farmers have started the threshing of whole plant of maize in wheat threshers which are commonly available. This resulted in breakage of grain from 25-30 per cent reducing the quality of grain produced and also the output of grain from thresher is reduced substantially in the range of 1.5 to 2 q/h but the practice is followed to fulfill the requirement of chaff (bhusa). In view of this a whole plant maize thresher has been developed by MPUAT Udaipur with the objectives to do the shelling of maize cob and simultaneously stalk is converted to chaff (Anonymous, 2010). Thresher has been designed for 5.5 kW power with spike tooth cylinder having bolts (6-7 per row). Concave is made by 8 mm square bars at a spacing of 18 mm (Fig. 8.31). Threshing speed is kept as 620 rpm. The output of machine is 170 kg/h. A tractor operated multicrop thresher has also been modified with similar arrangement. The output of grain is 540 kg/h with chaff size of 18 to 52 mm.

Fig. 8.31: Whole plant maize thresher

Corn harvesting machines

Four kinds of corn harvesting machines are in use i.e. corn snapper, corn picker, corn picker sheller and corn combine. Corn snapper removes or snaps the ear from the standing stalk but does not remove husks. Corn picker also called snapper husker snaps the ear from standing stalk and removes the husks. Corn picker sheller snaps off the ear, removes the husks and shells the kernels from the cobs all in one operation. A picker sheller attachment is available for the conventional corn picker. Corn combines have two, three, four, six or eight row heads as attachments (Fig. 8.32). These attachments replace the grain cutter bar. The corn header snap the ears from the standing stalks and feed the snapped ears into combines. Combine cylinder shells the kernels from the cobs and separate and clean the grain as they do in harvesting small grain. Standard corn heads harvest rows from 90 to 110 cm width. The 50 cm rows corn heads are available to harvest 4, 6 or 8 rows at a time. There are different units in a corn picker.

Gathering unit: Main parts of this unit are the snouts. These are hinged and can be set to float along the ground to slip under down stalks and then lift them into the gathering chains. Steel shoes beneath the snouts prevent them from digging into the ground.

Snapping unit: Ears are snapped from the stalks by a pair of long, closely spaced rolls. Because of their close spacing with upward plants and rotation towards each other, the rolls pull the stalks through quickly. The close spacing does not allow the ears to pass through instead they are passed on to husking rolls. Coarse trash and leaves are removed by ejectors and to husking rolls.

Husking unit: Husking units are available in two different types viz. continuous or combination rolls having the husking rolls as continuation of snapping rolls and separate husking unit has separate driven husking rolls independent of snapping rolls. In the both types one or more pairs of husking rolls are provided for each row of corn. Various roll designs are used in different models.

Cleaning unit: Mechanical trash ejector and air blast accomplish the cleaning. Ejecting rolls and augers remove trash and husks from the husking rolls and discharge it to the ground. The air blast from the fan strikes the ears as they leave the husking rolls. The strength of the air can be adjusted as need be, to blow of loose husks, leaves and trash etc.

Shelled corn saver: Some kernels may be shelled from the cob by the snapping rolls and some by the husking rolls. Kernels shelled off by snapping rolls fall to the ground and are lost, the kernels shelled off by the husking rolls are saved because they fall into the shelled corn-saver where they are cleared and then elevated to the wagon. It consists of a screen and a chain career.

Shelling attachment: Some models of corn pickers have shelling attachments. Revolving lugs on a screw feed cylinder rub the ears against one another and against the round bars of the cylinder cage to give the shelling action.

Wagon elevator and hopper: The hopper at the rear end of the machine receives the husked ears and also the shelled corn from the corn saver. The elevator can be controlled/adjusted for loading and unloading the wagon.

The grain combine owning farmers can use their combines for threshing of maize with husk (Gupta *et al.*, 1985). The adjustments for maize shelling needed are change of cylinder speed, concave clearance and adjustment of chaffer. The tractor-operated combine can thrash about 20 q/h of clean grain. The labour involved in dehusking is completely saved by the use of this machine.

Fig. 8.32: Self-propelled corn (maize) combines

Harvesting of sugarcane

Sugarcane harvesting is a highly labour intensive operation. Manual harvesting of sugarcane is a common practice. Different types of sugarcane harvesting knives of different sizes, shapes and weights are used at different places (Fig. 8.33). Labour requirement for sugarcane harvesting ranges between 800 – 1000 man-h/ha in tropical regions of India whereas in sub-tropics it ranges between 1200-1500 man-h/ha, which is highest as compared to other cultural operations. Sugarcane harvesting involves base cutting of sugarcane, stripping and de-trashing, de-topping, bundle making and finally transport to the sugar mills (Singh, 2014; Yadav, 2003). Delayed harvesting affects the quality of sugarcane yield, juice quality and sugar recovery. Scarcity of labour during peak agricultural operations necessitates introduction of sugarcane harvesters particularly in the states like Maharashtra, Tamil Nadu, Andhra Pradesh and Karnataka where harvesting is done by paid labourers.

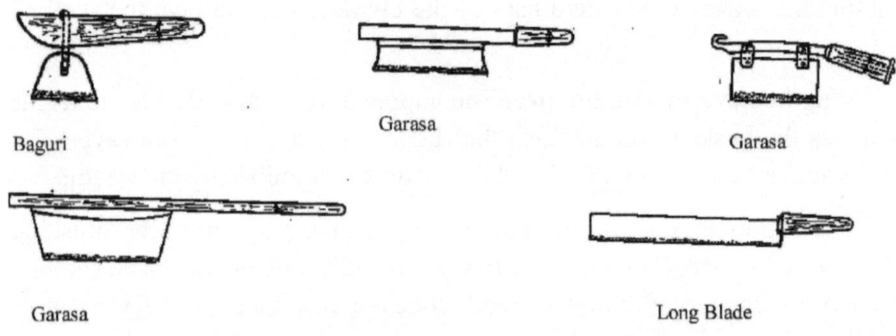

a) Varying shape and size of cutting knives.

b) Harvesting knife developed under NATP-ICAR

Fig. 8.33: Sugarcane harvesting knife

Sugarcane harvesters being used in India are of different makes and models. Whole talk sugarcane harvesters are quite suitable for those areas where sugarcane crushing is not possible within short period of harvesting. Delay in transport, loading, unloading, waiting at one or the other stage is unavoidable. Use of whole stalk sugarcane harvesters is also useful for those areas where green tops recovered are used for cattle feed. Such machines are either tractor operated or self propelled. Introduction of such machines in phases is possible even in those areas where sugarcane is not lodged; canes may be bent up to 10 - 15^0. There are only few designs of whole stalk harvesters in use in the world because of difficulty in handling of lodged cane by these machines. Therefore, whole stalk harvesters with base cutting, de topping and partial de trashing for erect, medium to low tonnage crop areas may be introduced, especially under those locations where self stripping / erect type cane varieties are common. Mechanical harvesters may be of bin type of 1.5 to 2 tonne capacity or whole stalk harvesters windrower type. Windrowing type harvesters may be backed up by grab loaders. These machines includes Case Austoft, Moller bin type, Cameco whole stalk, tractor operated Carib, Bonnel & Bunmai harvester windrower (Yadav, 2003). Presently production of this type of machine is almost stopped.

Different designs of sugarcane harvesters including tractor operated whole stalk, self propelled whole stalk and self-propelled billet type sugarcane combine harvester are available. It has been observed that these machines are designed for wide row spacing. Maintenance of these machines is a costly and difficult preposition. These designs do not fit well in the socio economic and agronomical practices. Based on the experience gained and observed data under Indian conditions, a self-propelled chopper type sugarcane combine harvester has been developed after deciding different factors for its suitability and agro techno-economical required (Fig. 8.34). The development of sugarcane combine harvester has been done jointly by NATP-ICAR, VSI-PUNE and M/s Rane Industrial Oven and Agro Machinery (RIOAM), Pune (Yadav, 2003). The first prototype of indigenous sugarcane chopper type harvester was developed during

VSI-MERADO Cane cutter VSI RANE Sugarcane Chopper harvester

Fig. 8.34: Field trials of indigenous chopper type sugarcane combine harvester

May 2003. The harvester was duly modified after field testing. Machine worked satisfactorily even under adverse conditions. In fact, for best performance of harvesting machine properly levelled, stone free and straight rows with minimum hurdles at optimum soil moisture conditions and crop conditions are desired.

Imported chopper type sugarcane harvesters mainly from New Holland are working in the field mostly concentrated in Tamil Nadu and Maharashtra. In this system, sugarcane is cut first from the top and then from the base and it is cut into billets and is loaded in transport cart. It does not require separate loader for loading purpose. The machine consists of gathering mechanism, detopping mechanism, base cutter, feed conveyor, chopper, and elevator and cleaning unit (Fig. 8.35). The machine is generally powered by the engine of 150 hp. Mostly it can harvest cane grown at a row-to-row distance of 90 cm. The average capacity of the machine is about 0.4 ha/h. It is most sophisticated machine among all types of sugarcane harvesters.

Ratoon management: Sugarcane crop is only profitable if two or three subsequent ratoon crops are taken after harvesting plant crop. In case of ratoon crop cost of land preparation, sugarcane seed and planting is saved. However gaps are filled up in case clumps are damaged by rodents and pests. It is very essential to harvest the sugarcane from the ground level for better ratoon crop and tillering. A tractor operated stubble shaver cum fertilizer applicator has been developed and commercialized.

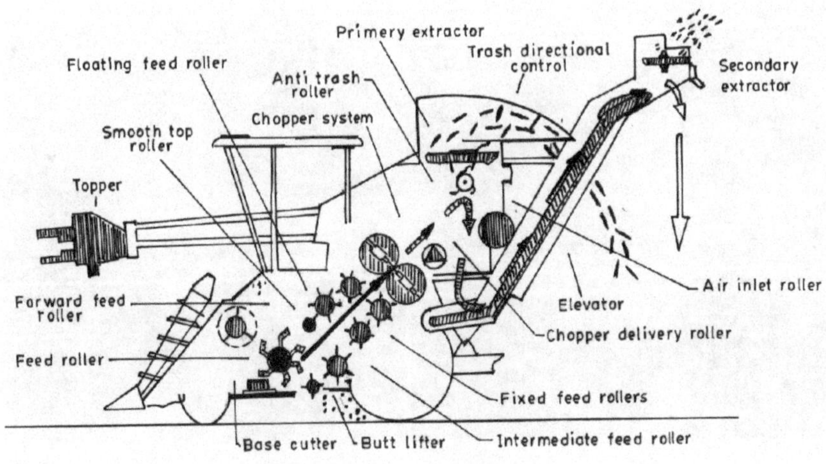

Fig. 8.35: Details of sugarcane combine.

Cotton harvesting/picking equipment

Manual harvesting/picking of cotton is quite a labour intensive operation. Mechanical harvesting of cotton is widely used in USA, USSR, Egypt etc. Mechanical pickers are best suited to irrigated areas and region of high rainfall (humid region) where yields are high, the fibers are long, the bolls are of open type and vegetable growth of plant is high. Strippers are most successful in dry areas (arid region) where plants are small (less vegetative growth) and have storm resistance bolls and fiber are rather short. The cotton-picking machines are area and plant specific. Thus one cannot be substituted for other.

Cotton is a perennial, indeterminate plant, and unlike most commercial crops produces flowers and fruit continuously during the growing season. Flowers and new bolls may be formed during the harvest phase, even though the bolls may never mature. Crop maturity extends over a considerable period of time. For best use of the mechanical picker, cotton plants should be erect, sturdy and of growth habits that permit penetration of insecticides and defoliants. Inner surfaces of the dry burs should be relatively smooth; burs should be straight and approximately equidistant from each other. These are inherent conditions that are often influenced by adverse environment. Where machine picking is contemplated, insect control must be timely and effective to insure good distribution of the crop over the plants. With well-balanced crops several known varieties stand erect with up to four bales per acre, but if out of balance may lodge badly with much lower yields. Lack of proper insect control is conductive to excess plant growth, severe lodging and other conditions which interfere with machine picking. Ideal cotton is one that will remain in the boll until harvest. The important factors to be considered in lock removal from the boll are: number of locks in a boll, branch number, node number bloom date, boll opening date and time interval between boll opening and boll picking. The other important factors affecting the retention of cotton in the carpel are carpel angle and lock angle. These may affect: (a) the force as the lock is removed, (b) distance between the point of lock attachment and the carpel during removal, and (c) integral of force with respect to time, which is converted to energy. Agronomic practices such as row spacing and plant spacing along with other agronomic practices like irrigation, spraying, fertilizer application etc are the major factors affecting the plant characteristics and viz. a viz. the mechanical harvesting.

Machines used for cotton harvesting are basically of two types i.e. stripper or picker type (Singh, 2007). Both these types of machines require the cotton variety with compact sympodial/semi-sympodial plants with synchronized boll opening. Mechanical pickers are selective as the seed cotton is removed from open bolls, whereas green, unopened bolls are left on the plant to mature for later picking. Stripper on the other hand is once over machines. All bolls whether open or closed are removed from the plant in a single pass. There are two types of cotton picker viz. drum type and chain belt type and these are of single row, double row and four row type. Basic functions of cotton picker are:

i) an arrangement for guiding the cotton plants with the picking zone and providing the necessary support while the seed cotton is being removed,

ii) devices to remove the cotton from the open bolls,

iii) a conveying system for the picked cotton, and

iv) a storage basket or container in which the picked cotton can be accumulated.

Two types of spindles are used: tapered spindles and small diameter straight spindles. Both types of spindles are moistened for cotton fiber adheres to the spindle so that lock can be wrapped around the spindle as it is removed from the boll. Working of cotton picker is:

i) The plants are folded into the throat of the machine by rounded members and two limbs and bolls are lifted up by limb lifting fingers.

ii) The revolving drum or belt of rotating spindles passes moistening pads and spindles are thoroughly wetted.

iii) The spindles then are projected in among the limbs and bolls to emerge the cotton. The plants are passed against the spindles by spring load pressure plates.

iv) The spindles rotate and come in contact with rubber doffers or stripper bar doffers.

v) The cotton drops into the air conveying system, which conveys the cotton to the basket.

vi) Air, dust and trash are exhausted throughout grates, which deflect the cotton into the basket.

vii) Hydraulic cylinders tilt the basket and dump the cotton into a trailer.

Drum type cotton picker

Drum type cotton picker is having two picking units and each unit has two picking drums. Each drum has the revolving spindles to remove the seed cotton from the plants (Anonymous, 2010). The picker is also having other important units such as doffer unit for removing seed cotton from the spindles, seed cotton conveying unit, storage basket and spindle moistening unit. Tapered spindles are used in drum type cotton picker. The spindles are mounted on a bar, the top end of which has a crank arm and a bearing that travels in a cam track (Fig. 8.36). The cam actuated picker bars swing the bars around so that the rows of spindles are about $1^{1/2}$ inch apart as they enter the cotton plants. The spindles are spaced $1^{1/2}$ inch both vertically and horizontally. The proper orientation of the spindle bars in relation to the row or to the doffers is obtained by means of a stationery cam and followers on the bars. All linear motion of spindles while in the picking zone is at right angle to the row. The spindles enter the cotton plant through a grid section, which subsequently prevents the plants from being pulled into the doffing area as the loaded spring withdraws. The spindle drums are operated in pairs, one drum on each side of row, but not directly opposite. The front and rear

Fig. 8.36: Tapered spindles used in drum type cotton picker
Courtesy: New Holland Fiat India Pvt. Ltd.

a) Tractor operated two row spindle type cotton picker
Courtesy: John Deere India Pvt. Ltd.

b) Self-propelled two row spindle type cotton picker
Courtesy: New Holland Fiat India Pvt. Ltd.

Fig. 8.37: Drum type cotton picker

drums have different speeds 60 rpm and 79 rpm. The spindles also rotate at different speeds (2200 rpm and 2700 rpm) at the same forward speed (generally 3.2 to 5.6 km/h) as the machine moves along the row (Fig. 8.37), Anonymous, 2010. High drum has 20 spindles per bar and low drum model has 14 spindles per bar.

Chain belt type spindle picker

The chain belt spindle is an arrangement so called because the spindle bars are attached to an endless chain belt. The belt consists of 80 spindle bars or slats each containing 16 almost smooth spindles. The belt is geared to travel at about (4.8 km/h) which is also low travel. Generally the spindles project into the row of plants only from one side. If two belts of spindles are arranged in tandem on opposite side of the plant, row spindles projects onto the plants from each side of plant row.

Cotton strippers

Mechanical stripping of cotton is accomplished by forcing the plants through an area too small for the bolls to pass. When plant is passed through the machine head, forward motion of machine applies a force in a direction perpendicular to ground, which separates the cotton bolls from plant. These snapped bolls are collected in the machine while plant remains in the row. The roots of plant should be strong enough to withstand the force applied by machine. There are two types of cotton stripper viz. finger type and brush type (Singh, 2007). The finger on the comb type stripper is 61.0 to 91.4 cm in length and set at angle of about 10 degree to 15 degree to the horizontal. Spacing between fingers is kept 16 mm. It consist of stripping fingers, stalk walker kicker, auger, cleaning grids, kicker wheels, green boll separator, pneumatic air conveyor system, grate and auger to pneumatic conveyor system. As the finger type strippers moves through the field, the forward motion of the inclined teeth snapped the cotton bolls from the plant. Stripped cotton is moved up in the inclined teeth by the next group of stalks being harvested. A reel mounted over the rear of the teeth kicks the cotton into a cross auger. Then the cotton passes through the cleaner which utilizes the air velocity method and combines cleaning techniques of combing and sling off type to remove the sticks, burs, crushed leaves and soil from the mature open bolls. The clean seed cotton is then moves pneumatically up into the basket. Finger type strippers work well only under rather limited conditions; such as plant should be small having low vegetative growth, properly defoliated and atmosphere should not be damp (Humid). Operating speed usually does not exceed 4.00 km/h and sometimes must be considerable lower. To counter the limitation of finger type stripper, brush type stripper has been introduced. The advantage of finger type stripper is that it can be used for harvesting of any row width of cotton.

Brush type cotton stripper is classified according to the number of rows it can harvest and type of brush. Brush type stripper employs a pair of rolls that are approximately 1020 mm or 1170 mm long and 152 to 165 mm in diameter and are mounted at an angle of about 30 degree above horizontal. Their surfaces are fitted with longitudinal brushes alternate with rubber paddles at 45-degree interval around steel tube (behind). The rolls are driven at about 600 rpm with their adjacent surfaces moving upward next to the plants. As the bolls are snapped off, they are moved away from the plants by the surface of rolls and delivered on to the adjacent conveyors. These conveyors convey seed cotton to cleaner, which cleans the trash content from seed cotton, and then clean seed cotton is conveyed into basket. The clearance between rolls is adjustable and usually set at 3 to 6 mm. Center to center spacing of two rolls is 1.02 m. The advantage is that the flexibility of the roll surfaces permits a degree of automatic adjustment for light and heavy growth of plant. The spacing between the rolls can be adjusted manually to meet extreme conditions. It has limitation that at high rates of travel, it is difficult to steer the tractor so that narrow space between the stripping rolls is always in line with the row of plants.

Two row brush type stripper has been developed to meet the demand of growing cotton in narrow rows having spacing 30 to 40 cm apart. Two roll units have a center to centre spacing of 356 mm instead of 1020 mm in single row stripper. This is the minimum practical spacing with the 165 mm diameter rolls. With careful driving double row spacing of 30 cm to 40 cm on each bed can be accommodated satisfactorily. Part of the stripped cotton from both rows passes down through the 25 mm space between the two center rolls into an auger conveyer that required a tapered front section to fit into the limited space available between the rolls and top of bed. Cotton moved outward by the rolls is collected by two straight augers located below and outside of the outer rolls and conveyed rearward to the cross auger. The rolls and the row unit augers are operated at 600 rpm. Factors which made the cotton stripper more popular than cotton picker are:

- Lower initial cost
- Lower maintenance cost

- Development of improved cotton varieties (early maturing & storm proof bolls) better adapted to stripping
- Improved strippers employing feasible rolls, green boll separation and cleaner fox trash content removal.
- Improved ginning equipment for separating trash
- Trend toward closer row spacing, which reduces vegetative growth of plant.
- Good recovery of cotton in the field
- Higher harvesting speed.

Application of defoliants: For dependable defoliation, fields should be evenly grown and all conditions, such as unleveled land and salinity, should be corrected before the crop is planted. Fertilizers should be judiciously used and nitrogen in particular should be fairly well depleted before maturity. The harvest should follow close behind abscission of the leaves. Defoliants reduce the harvest season and increase the percentage of leaf drop depending upon the mode of defoliant application. Early defoliation increases the quantity of high-grade cotton which may be taken in first pick. Late defoliation increases losses in the first pick over those from green picking.

Threshing equipment

The operation of detaching the grains from the ear head, cob or pod is called threshing. It is basically the removal of grains from the plant by striking, treading or rupturing (Fig. 8.38). The traditional method of threshing using manual labours requires 150-230 man-h/ha. Threshing is normally done after the grain moisture content is reduced to 15 to 17%. In various parts of world, threshing is accomplished by treading the grains under the feet of animals or under the tractor tyres, striking the grains with sticks, pegs or loops and removing the grains by rubbing between stone or wooden rollers on a threshing floor or between the rasp bar and a concave of combine. The threshing can be achieved by three methods: rubbing action, impact and stripping. Traditional threshing is done by trampling of paddy under feet, beating shelves of rice or wheat crop on hard slant surface, beating crop with a flail, treading a layer of 15 to 20 cm thick harvested crop by a team of animals. Threshing by bullock treading is practiced on large scale in the country.

Fig. 8.38: Manual threshing by beating on the ground surface or wooden log

Thresher is a machine to separate grains from the harvested crop and provide clean grain without much loss and damage (Singh, 2007; Singh and Verma, 2009; Pandey *et al.*, 1997; Banga *et al.*, 1984; Mehta and Sharma, 1985 and 1986). During threshing, grain loss in terms of broken grain, unthreshed grain, blown grain, spilled grain etc should be minimum. Bureau of Indian Standards has specified that the total grain loss should not be more than 5 per cent, in which broken grain should be less than 2 per cent. Threshers are the most important component of farm mechanization. If threshing is not done timely, all efforts made by farmers and inputs given to crop goes wasted. Traditional method of threshing by animal is very slow. It gives low output. Due to low output, the cost of operation is high and there is a huge loss of grains because of rodents, birds, insects, wind, and untimely rain and fire hazards. Wheat threshers overcome these difficulties to a great extent. Wheat threshers are of two type viz. animal-drawn and power threshers. In animal-drawn threshers, olpad thresher is a common machine used in different parts of the country. There are mainly five types of power threshers in use namely, drummy thresher, power wheat thresher, spike tooth/peg tooth type thresher, chaff cutter/ syndicator type thresher and multicrop thresher. Power wheat thresher is a machine, which thresh the wheat crop and performs several other functions such as feed the harvest crop to the threshing cylinder, thresh the grain out of the ear head, separate the grain from the straw, clean the grain, and make 'bhusa' suitable of animal feeding.

The famous Ludhiana thresher was first introduced in India during 1956-57 (Mehta and Sharma, 1985). The thresher was tractor operated type and used mainly for wheat. It was a very good machine, which threshed, cleaned and bagged the grain; at the same time it made the quality straw (bhusa). Further development work took place during the period from 1965 onwards for low horsepower threshers. The most widely used design; spike tooth cylinder thresher was commercially marketed in the country around 1970 (Manes *et al.*, 1995). This simple design has been able to maintain the cost of machine low as the total weight of machine is greatly reduced. The output capacity also improved. These threshers are available in various sizes operated by 3-40 hp power sources. The grain output is 20-25 kg/hp-h. Beater type threshers take comparatively more power than spike tooth threshers.

Working principle of a thresher: During operation, the crop material is pushed into the threshing cylinder through the feeding chute, which gets into the working slit created between the circumference of the revolving drum having attached spikes and the upper casing. The material layer is struck several times by the spikes against the ribs, causing threshing of the major amount of grains and breaking stalks into pieces, and also accelerating them into the inlet of the lower concave.

Functional components of threshing unit

A power thresher essentially consists of feeding unit, threshing unit, cleaning unit, power transmission unit, main frame and transport unit (Fig. 8.39). The operation of conveying the cut crop into threshing unit is known has feeding. Normally, one of the two types of feeding units 'throw-in-type' or 'hold-on-type' is used in power threshers (Fig. 8.40). In 'throw-in-type' feeding unit, the cut crop is pushed into threshing cylinder, where as in 'hold-on-type' the heads is only pushed into the cylinder and straw is manually or mechanically held. Throw-in-type feeding device is quite common in the threshers, which may be a feeding hopper or feeding chute. In feeding hopper type of feeding device there is a hopper, placed on the top of the threshing cylinder. Generally hopper type of feeding units have a rotating star wheel mechanism between the hopper and threshing drum to facilitate the uniform feeding of crop to the drum. The initial cost of this system is high, hence is mostly used on a large thresher e.g. axial flow thresher of large capacity. The threshers are generally of spike-tooth cylinder type, chaff-cutter type and hammer mil or beater type. The crop is fed manually into these machines through a feeding chute. It has been observed that human factors such as inattentiveness, unskilfulness, overwork, and physical incapability, wearing of loose clothes, hand-wears and use of intoxicants are mainly responsible for about 73% of accidents (Verma *et al.*, 1978). The machine factors such as improper design of feeding systems, substandard material and defective design contribute to about 13% of accidents. Crop factors such as feeding of ear-heads, short crop stalks and wet crop contributed about 9% of accidents whereas inadequate light, crowded surroundings and slipping in the threshing yard contributed to about 5% of accidents. The threshing accidents can be minimized provided the farmers adopt the following measures:

- The farmer should buy only those threshers, which are fitted with safe feeding chute as per B.I.S. standards (Fig. 8.41). For safety, the minimum length of feeding chute should be kept 90 cm, covered up to a minimum of 45 cm and inclined to the horizontal at an angle of 5-10 degrees. The angle of covered portion with the base length of feeding chute should be kept equal to 5 degrees.
- Employ only skilled and trained workers for feeding the crop to the thresher. Avoid feeding ear-heads without stalks as it may lead to serious hand injuries. While feeding the crop or ear heads, operator should not insert the hand(s) deep in the feeding trough. The thresher should be operated at the speed recommended by the manufacturer or Bureau of Indian Standards (BIS), Table 8.3. Similarly, feeding of wet crop should also be avoided which otherwise might lead to fire accidents.
- Avoid talking while working on the thresher. Do not work on thresher under the influence of alcohol or any other intoxicants. Do not work on thresher for more than 8 hours or when feel tired. Do not wear loose clothes, wristwatch, and bangles while working on the thresher.
- Ensure proper lighting in case the machine is to be operated at night, otherwise poor visibility may lead to accidents.
- Keep the work place and surroundings of thresher free of all kinds of obstructions.
- Do not smoke or light a fire near the threshing yard.
- Do not cross over the flat belt or move near it.
- Keep a first aid box handy for use in the event of need.

Table 8.3: Recommended speeds of threshing drum for different crops

Crop	Speed of threshing drum	
	rpm	m/s
Paddy	675-1000	16-25
Wheat	550-1100	20-30
Barley	740-1080	20-26
Gram	400-750	12-22
Jawar	400-675	12-20
Bajra	400-550	12-16
Peas	430-750	12-22

Source: IS 9019: 1979

To achieve the desired speed of threshing drum, the size of thresher pulley can be computed by

$$Dt = \frac{Dp\ Np}{Nt}$$

Where,

Dt = diameter of thresher pulley, mm

Dp = diameter of prime mover pulley, mm

Nt = speed of threshing drum, rpm

Np = speed of prime mover pulley, rpm

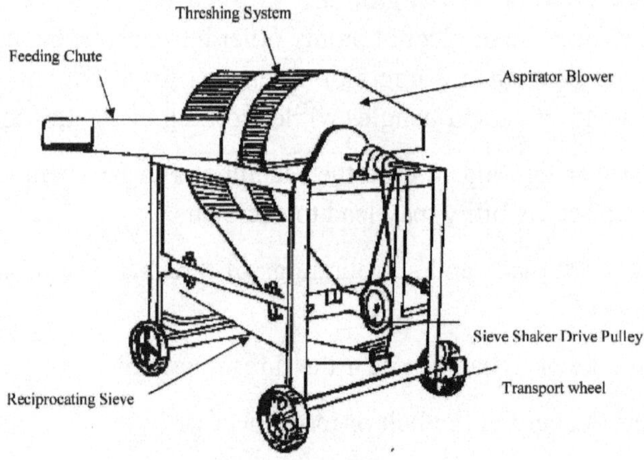

Fig. 8.39: Details of wheat thresher

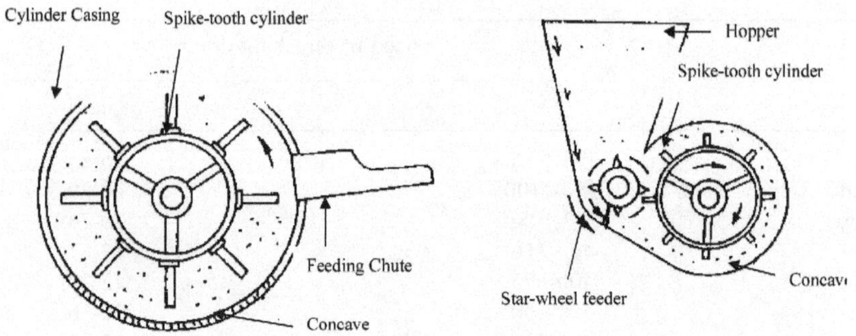

Fig. 8.40: Different types of crop feeding systems.

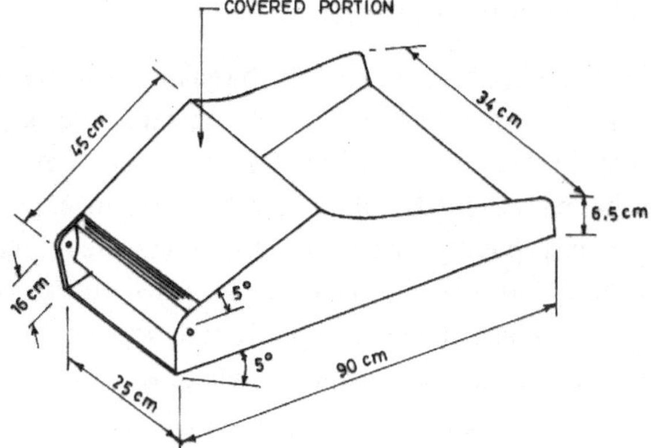

Fig. 8.41: Specifications of BIS recommended safe feeding chute.

Threshing unit

The threshing is accomplished by the impact of the rotating pegs mounted on the cylinder, over to the ear heads, which force out the grain from the sheath holding it. In the threshing of wheat crop, the straw is also bruised and broken up by the impact, thus converting it into 'bhusa' (straw). Threshing unit is mainly consists of a cylinder and concave. There are different types of threshing cylinders (Fig. 8.42) such as spike tooth/peg type cylinder, rasp bar type cylinder, angled bar type cylinder, wire loop type cylinder, cutter blade or syndicator type cylinder and hammer mill type cylinder.

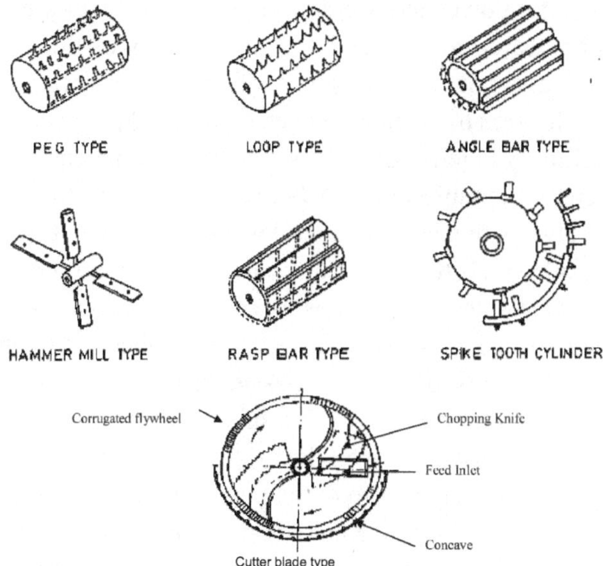

Fig. 8.42: Different types of threshing cylinders.

Spike-tooth threshing system

Spike-tooth threshing system typically has a threshing cylinder fitted with spikes which may be flat or threaded bolts to facilitate adjustment of the cylinder-concave clearance. The casing of the cylinder comprises of the upper concave, which has bars welded to blind sheet whereas lower part of the concave comprises of a grate. Typically the concave grate comprises square bares welded at a spacing of about 8 mm. Axial spacing of grid is about 50 mm. Direction of the threshing cylinder is such that the crop materials after entry into the threshing system impacts against the upper concave where these are threshed/bruised. Subsequently, these materials are passed through the concave opening in lower concave after being reduced sufficiently. In the process, grains are thrashed and separated and the straw is reduced suitably to pass through the concave, because the crop materials receive repeated impacts by the threshing cylinder against the concave. There are several variations in the design of spike tooth threshing system. These variations for the same range of horsepower include cylinder diameter, number of spike rows, number of threshing spikes per row, diameter of spikes, shape of spikes etc. Besides these variations, the effect of operational variables like feed rate, peripheral speed and cylinder-concave clearance would not be exactly the same as for the conventional rasp bar/spike tooth threshing system. The specific power requirement (kWh/q) initially decreases with increase in feed rate up to a certain minimum level beyond which the specific power requirements start increasing (Manes *et al.*, 1995). Increase in specific power requirements beyond this level are due to higher feed rates as compared to optimal handling capacity of the given threshing system. Higher values of specific power at feed rates below this level are due to inefficient utilization of the input power. Grain crackage remains unaffected, whereas, threshing efficiency decreases slightly with increase in feed rate but is always within acceptable limits (>99.5%). In fact threshing efficiency for such a system always reached a desired level irrespective of the variations in the system parameters, if the concave opening size is not more than 8 mm and peripheral speed is not less than 20 m/s. Variation of peripheral speed on either side of this optimal level increases the power requirements. At lower peripheral speed, the impact level is low, which does not bruise the straw effectively and thus the crop retention time within the threshing system results in crop receiving more impacts of relatively lower level, which results in higher power requirements. Increase in power requirements at peripheral speeds beyond the optimum level is mainly due to the increase in the no-load power requirements. Grain crackage and threshing efficiency increase with increase in peripheral speed.

Bar to bar spacing in the concave is perhaps the most critical design factor, which affects the power requirements, grain crackage and straw size index. Threshing efficiency can also be affected if the bar to bar spacing in the concave is increased beyond a certain level (10 mm). With the increase in bar to bar spacing, grain crackage is bound to decrease and the straw becomes coarser resulting in higher values of straw index. Total power requirements are lower at higher bar to bar spacing in the concave, as the straw does not have to be reduced to finer size. The spacing between louvers/bars in the upper concave casing does not affect power requirements, grain crackage and threshing efficiency but results in lower values of straw size index (finer straw) at closer spacing between these louvers/bars. At closer spacing, the number of bars is more, which results in relatively finer straw. Moreover, the crop handled by these threshing systems is dry and thus is easy to thrash.

Shape of the spikes has a profound effect on the power requirements, grain crackage and the straw size index. Flat shaped spikes require minimum power even when these result in finer straw than round/square shape spikes. Grain crackage is minimum in the case of round shape spikes. Low power requirements in case of flat shaped spikes may be due to the fact that front edge of the spike is relatively sharp, which is more efficient in bruising the straw. Some manufacturers prefer to use flat shaped spikes instead of round spikes especially when the thresher is to be used for wheat only. Round spikes are preferred under the conditions where a variety of crops have to be handled because in such cases adjustment of cylinder-concave clearance is required, which is easier in the case of round spikes. Increasing number of rows of the spikes results in decreasing the power requirements. When the number of rows are more, straw is bruised relatively faster which escapes through the concave. Thus, straw bruising is more efficient, if the number of rows is more. Increasing the number of rows, however, does not result in any effect on the threshing efficiency, grain crackage and straw size index.

Length of spikes for the same diameter of the cylinder does not affect power requirements, threshing efficiency and straw size index. However, it affects grain crackage i.e. lower values of grain crackage are observed when length of spikes is higher. With increase in diameter of the spikes, grain crackage increases whereas straw size index decreases. Thus, it appears that with the increase in diameter of the spikes, frontal area of the spikes increases, which imparts impact to more crop materials that results in higher values of grain crackage and causes straw to be finer. Power requirements, generally, does not increase with increase in the diameter of the spikes. Higher values of power requirements at 15 mm diameter of the spikes are perhaps more because the number of spikes per row has to be increased to keep the same end to end axial spacing between the spikes. Increase in axial distance between the spikes (i.e. lesser number of spikes per row) results in lower values of grain crackage and higher values of straw size index (i.e. coarser straw). When the number of spikes per row is high, power requirements are high because of the greater resistance offered to the movement of the threshing cylinder, whereas at lower values of number of spikes per row, crop retention time within the threshing system perhaps increases.

Chaff-cutter type threshing system

A chaff cutter type thresher can handle crop with grain moisture content up to 24.5% wet basis with reduced output (43-52%), high grain crackage (about 10%) and relatively inferior quality of 'bhusa' (split straw percentage of about 75% as compared to about 95% for dry crop). In view of the advantages of lower energy requirements in case of chaff-cutter type threshers, high capacity threshers for wheat based on this system called 'HARAMBHA' have become very popular in northern Indian states particularly Punjab and Haryana (Mehta and Sharma, 1985 and 1986). In the existing design of chaff-cutter type threshers, materials after cutting, which contain grains, unthreshed heads, nodes and 'bhusa' have to pass through the threshing mechanism (Fig. 8.43). Ideally, it may be possible to separate 'bhusa' just after cutting. This reduces the materials handling by the threshing system by 49 - 57% and results in reduction of power requirements. It consists of 2-4 chaff cutter blades mounted on the radial arms of a flywheel. The rim of flywheel is corrugated to provide additional rubbing action. This type of cylinder performs satisfactorily in moist crops. Moist crops, in case of spike tooth or hammer mill type cylinders results in frequent choking, over loading etc. This type of thresher is generally provided with position feed roller and endless conveyor belt. This is essentially an adoption of chaff cutter for threshing. The crop is fed as is done in case of chaff cutters. After passing through a set of rollers, crop is cut into pieces. Varying the set of gears can vary the size. Three to four serrated blades are fastened on the radial arm of the flywheel. Threshing is done mainly due to cutting helped by rubbing and impact. However, chopping knives need to be sharpened every 3-5 hours of operation. The machine is more prone to accidents due to positive feed rollers.

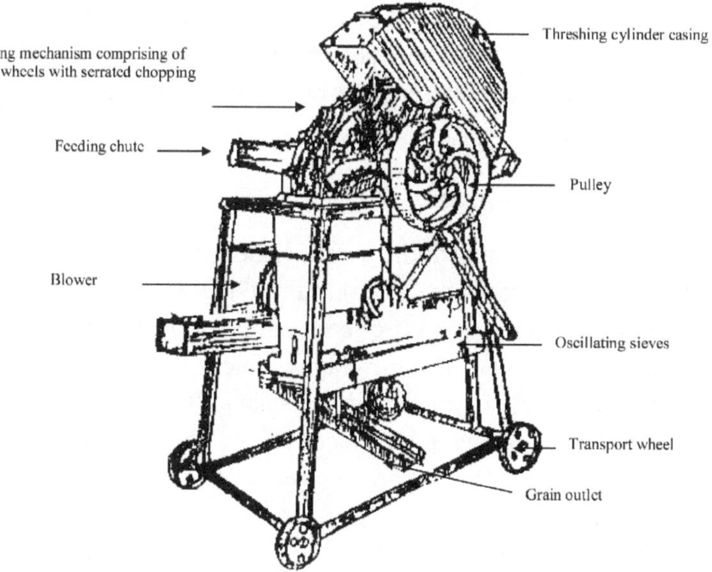

Fig. 8.43: Detail components of chaff cutter type thresher.

Serrated-tooth type threshing system

Serrated-tooth system has been incorporated recently in straw combines in view of their claimed lower power requirements. Attempt has also been made to incorporate this system in grain combines for wheat with straw bruising as an integral feature. In this system threshing cylinder uses 9 rows, out of which 6 rows has 3 serrated tooth strips and one spike, while the rest of the rows has four spikes per row. Thus, in all there are 21 serrated tooth strips and 12 spikes mounted on a threshing cylinder of 450 mm diameter. The threshing cylinder operating at 23.33 m/s with a concave bar spacing of 10 mm gives desirable levels of threshing performance (threshing efficiency > 99%, grain crackage less than 2%, size of straw 19.8 mm) at feed rate of 650 kg/h. The threshing performance in terms of threshing efficiency, grain crackage and straw quality, serrated tooth type system requires 12.3 to 29.6% less power as compared to the spike tooth type system within the feed rate range of 450 to 750 kg/h.

Rasp bar type cylinder

In this type of cylinder, there are slotted plates, which are fitted over to the cylinder rings, in such a way that the direction of slot of one plate is opposite to another plate. Corrugated bars are mounted axially on the periphery of the cylinder. It is fitted with an upper casing and an open type concave at the bottom of the cylinder. This type of cylinder is commonly used in threshers. It gives better quality of bhusa and it can be used for a wide variety of crops viz.- wheat, paddy, maize, soybean etc. The cleaning system is provided with blower fan and straw walker.

Wire loop type cylinder

In this type of threshing drum, there is hallow cylinder, over which a number of wooden or MS plates are fitted. On these plates, number of wire loops is fixed for threshing purposes. Wire-loops are fitted on the periphery of a closed type cylinder and woven wire mesh type concave is provided at the bottom. This type of cylinder is common in the manually operated paddy threshers. Holding the bundle against the loops of revolving cylinder does threshing of paddy crop.

Hammer mill type cylinder

It uses beaters to do the required job of threshing. The shape of this type of cylinder is different from the above-discussed cylinders. The beaters are made of flat iron pieces and are fixed radially on the rotor shaft. They are also called drummy type and beaters mounted on a shaft rotate inside a closed casing and concave. Generally feeding chutes are used with hammer mill type threshing cylinders. The cut crop is fed perpendicular to the direction of motion of rotating beaters. This type of thresher requires more power as compared to spike tooth type of thresher. It is provided with aspirator type blower and sieve shaker assembly for cleaning grains.

Axial flow type: It consists of spike tooth cylinder, woven-wire mesh concave and upper casing provided with helical louvers.

Angle bar type: In this case bar are welded on the periphery of cylinder at certain angle.

Drummy type, hammer mill type and syndicator type threshers are suitable for threshing wheat crops only and they can produce fine quality of '*bhusa*', rasp-bar type, wire-loop type and axial flow type threshers are suitable for paddy and they do not make fine straw. Rasp-bar type threshers can be used for threshing other crops but farmers do not prefer, this machine because it does not make fine '*bhusa*'; and cost is very high due to its bulky size. Though the hammer mill type threshers can produce fine quality '*bhusa*' its use is decreasing day by day due to high power requirement. Portable wire loop type paddle operated threshers are widely used by farmers in paddy growing areas.

Spike tooth type thresher can thresh wheat crop and can produce fine quality of '*bhusa*'. This thresher can be used for threshing other crops if the blower is mounted on a separate shaft so that the cylinder speed can be varied independently. Majority of farmers prefer spike tooth type threshers because of their simplicity in design, low cost and their ability to make fine '*bhusa*'.

Concave: Cylinder and concave together makes the threshing unit. It separates the grain from the crop and removes grain from the straw. Concave is provided in the thresher to hold the fed crop inside the threshing chamber and allows only grain and small amount of chaff to pass through it. The threshing takes place only in this space. It is a curved unit, made of iron steel or iron bar, fitted near the threshing cylinder. The clearance between cylinder and concave is adjustable, depending upon the size and type of grain. The concave clearance for wheat is 5 to 13 mm and for paddy is 5 to 10 mm. As the concave clearance is reduced, the threshing efficiency increases but losses increase and vice versa. The concave clearance at the inlet is less as compared to outlet. There are different types of concave, which are used in thresher.

Screen type concave: It is made of MS rod. It is semicircular in shape and sometimes made with wire also. The screen allows the material after threshing to pass through its perforation.

Perforated concave: In this, perforations are made in a mild steel sheet. The concave is closed from both the ends by iron sheet. The size of perforation is made as per the size of grain of a crop.

Cleaning unit: This unit is provided to separate the grain from chaff. It further uses sub units, like aspirators or blowers, sieves and sieve shaking mechanisms to separate out grains from chaff. The thresher that is provided with aspirator unit is usually called aspirator type thresher. Those threshers fitted with blower which blows air in horizontal direction is called drummy threshers.

Blower or aspirator: After threshing unit carries out threshing, the cleaning and separation of straw from grain is required. The fan is generally installed on the main shaft over which cylinders, flywheel and driven pulley are mounted. Fan lifts/sucks the lighter material chaff and other plant portion and throw away from the out let. Rest of the separation-cum-cleaning is done by screen with its oscillating motion.

Screens: Most of the power threshers are equipped with two screens. Top screen is provided so as to pass the grain to second screen and chaff etc is taken out from it. Other screen sieves out the smaller grain or weeds seeds and delivers the cleaned grain towards outlet. The size of screen hole is selected on the basis of grain size. These screens are effective when kept under oscillation.

Shaking mechanism: The screens are oscillated or shaken with a crank attached to the screen. This crank is powered from main axle either by belt or by rod. The circular motion of the main shaft is converted into oscillating motion of screen, which shakes it and separates the grain from other foreign material and chaff. The separating effectiveness depends on the frequency of strokes of crank, which is adjustable.

Power transmission unit: Threshers are usually powered with tractors and sometimes with electric motors or diesel engine also. After installing the thresher into the threshing floor in the field, tractor PTO shaft is coupled with a flat pulley. A corresponding matching pulley of appropriate size is provided over to the thresher main shaft. These pulleys are connected with a proper rating of flat belt and thresher is operated. Blower fan is provided into the main shaft of the thresher, which rotates and does the required job. The screens are oscillated with the help of a V-belt and a crank wheel, poweres with main shaft of thresher. A heavy flywheel is also provided on the main axle of the thresher. It is very important part of any thresher. It is provided to store the energy to supply continuously and equally to the entire threshing cylinder. It is made up of cost iron, and fitted on one end of the main shaft of thresher.

A very strong frame is provided in the thresher on which all the functional parts are attached. The frame is made usually of heavy angle iron sections. It should be strong enough to sustain vibrations of machine, during its operation in the field. Thresher is provided with wheels at its legs, so that transportation can be done easily. These wheels are made mostly with cost iron but new and large capacity threshers are equipped with pneumatic wheels for better performance during transportation.

Thresher adjustments: The following adjustments can be done on a stationary power thresher:

Cylinder and concave clearance: In order to get cleaned grains and proper threshing, it is very important to set the proper clearance between tip of cylinder and concave. On an average, concave clearance is kept about 25 mm at the mouth, 10 mm at the middle and 15 mm at the rear end. Start operating the thresher, by keeping proper recommended speed, and check if any grain is left in the ears. If it is so, reduce the concave clearance gradually, until drum is threshing cleanly. Too close concave setting is likely to crack some of the grains.

Cylinder speed: The drum of the thresher should be rotated at proper speed for better threshing and cleaning efficiency. Normally, manufacturers specify the cylinder speed for different crops. The cylinder speed can be checked using tachometer. Operator should check the speed occasionally under load for proper functioning of thresher. The cylinder peripheral speed for wheat is kept between 1520 to 1830 m/min and for paddy between 370 to 920 m/min.

Fan adjustment: Fan(s) fitted on thresher must provide the proper amount of blast. The shutter(s) at each end of fan should be adjusted properly so that it could provide blast sufficient enough to remove chaff and light materials without grain. Watching the sample and adjusting the blast can help in getting the desired results.

Drummy thresher

These threshers were very popular in the beginning when threshers were introduced because of its simplicity and low cost. The radially arranged arms known as beaters are mounted on the shaft (Fig. 8.44). These are made of mild steel square section with mild steel flat welded or bolted at the top. The beaters revolve inside an enclosed casing. Ribs are provided inside of upper half of the cover in order to have better threshing. The lower half (known as concave) has rectangular openings made out from square bars. The crop is fed through feeding chute. Crop receives impacts from the rotating beaters till size is reduced to pass through concave. The clearance between beater and concave is kept about 18-20 mm. The crop should be well dried before feeding in the

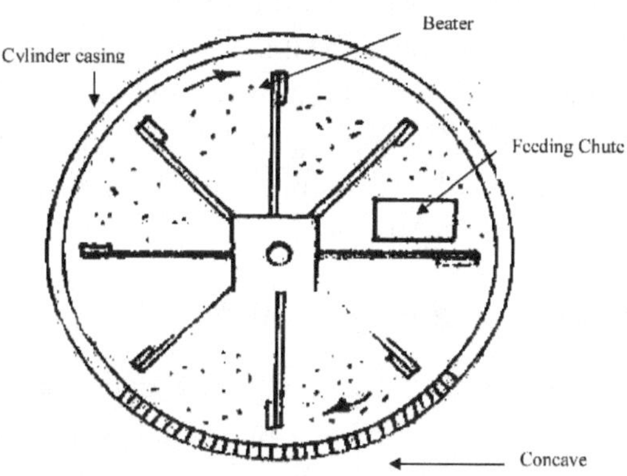

Fig. 8.44: Beater type drummy thresher.

thresher. A wet crop raps around the beater shaft and machine becomes overloaded. These threshers do not have provision for separation and cleaning of grains. The threshed material is later separated and cleaned by small pedal type blower.

Olpad thresher

'Olpad' threshers (Fig. 8.45) are also used for threshing wheat crop. A pair of bullocks pulls it around over the dried crop spread in a circular form on the threshing ground. Threshing is continued till the entire material becomes a homogeneous mixture of grain and '*bhusa*' (chaff). It consists of about 20 circular grooved discs each of 45 cm diameter and 3 mm thickness placed 15 cm apart in three rows. An operator's seat is provided on the frame to control the movement of animals. All discs are mounted staggered to give more effective cutting of the straw. It has 3 or 4 wheels to facilitate its movement from one place to other. Threshing by this thresher is fairly efficient and cheap but is quite slow with low output capacity. This machine can be used for threshing wheat, barley, gram etc.

Fig. 8.45: Olpad thresher

Pedal operated paddy threshers

Paddy thresher of pedal operated type (Fig. 8.46) consists of mainly a well-balanced cylinder with a series of wire loops fixed on wooden slates. It has got gear drive mechanism to transmit power. While cylinder is kept in rotary motion at high speed, the paddy bundles of suitable sizes are applied to the teeth. The grains are separated by combining as well as by hammering action of threshing teeth. It can also be powered by 0.5 hp single phase electric motor. Two persons can work at a time. It is simple, strong and gives quality service for longer period. Grain separation efficiency is about 90%. It can thresh about 200 kg/h. It is very easy to operate, safe and women friendly. It is low maintenance machine.

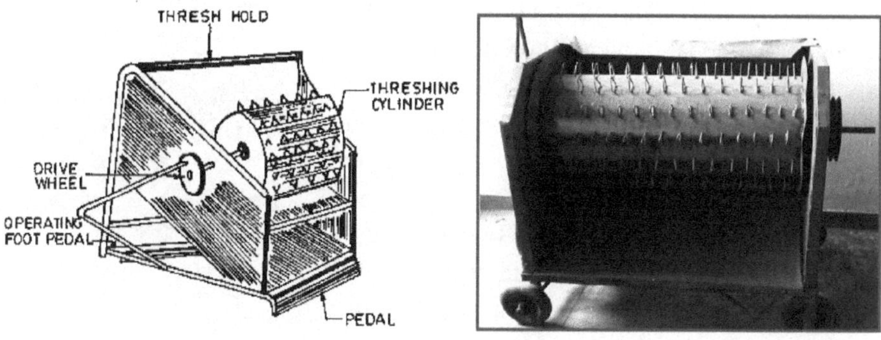

Fig. 8.46: Pedal operated paddy thresher.

Flow-through rice thresher

Fig. 8.47: Flow through rice thresher

It is a portable, straight flow type thresher having rasp bar cylinder, special concave, double sieves, blower and three straw walkers suitable for threshing green and moist paddy crop (Fig. 8.47). It has one centrifugal blower and powered by 8 hp engine/7.5 hp motor (Anonymous, 2009). It is operated at 920 rpm and gives 99% threshing efficiency and 97% cleaning efficiency. The output capacity is 1620 kg/h. It saves 88 per cent labour, 70 per cent operation time and 52 per cent cost of operation compared to conventional method of foot trampling and winnowing.

Portable paddy thresher

Axial flow power paddy threshers are also available in which crop is fed to threshing chamber through feeding chute (Anonymous, 2009). The thresher consists of a peg tooth cylinder with straw throwing paddles on one end enclosed by a cover with spiral louvers, wire mesh concave at the bottom and a blower (Fig. 8.48). The paddy crop is fed through the feed inlet. Paddy is threshed due to impact and rubbing action between threshing drawn loops and concave screen. The grain separation and axial movement of the straw takes place in threshing cylinder. The bruised straw is thrown from the straw outlet at the end of the threshing cylinder. The louvers provided on the casing help in axial movement of the straw. The grains are cleaned with the help of a fan and cleaned grain goes down through the grain outlet at the bottom of the thresher. They are available in different horse power range. It can be shifted from one field to another by two persons. Threshing efficiency is 98%, cleaning efficiency 90%, output capacity 100 kg/h and labour requirement 2 man-h/q.

Fig. 8.48: Portable rice thresher

Multicrop plot thresher

It consists of a spike tooth cylinder, concave, aspirator blower and sieve shaker (Fig. 8.49), Anonymous, 2009. Two top covers, three concaves, three sieves, variable cylinder speeds (7-21 m/s) are provided for threshing different crops. It is useful for precision threshing of samples in field experiments. The handling capacity of this equipment is 3-4 times faster over manual method of threshing crop samples. Its grain recovery is over 99.5 per cent. It is suitable for threshing wheat, maize, sorghum, rice, gram, pigeon pea, soybean, mustard, sunflower, and safflower and linseed crops.

Fig. 8.49: Multicrop plot thresher

High capacity multi-crop thresher

It is basically a chaff-cutter type thresher also called 'Harambha' thresher (Singh and Pandey, 2008; Garg and Singh, 2002; Pandey *et al.*, 1997). It consists of a threshing cylinder, concave, two aspirator blowers, reciprocating sieves, feeding chute, feeding conveyor, feed rollers, safety lever in the feeding chute and flywheel (Fig. 8.50). A platform is attached to the main frame of thresher, on which a person stands and feeds the crop into thresher. All the crop materials are fed through the conveyor of feeding chute and feed rollers move the crop into threshing cylinder. A safety lever provided in feeding chute prevents the entrapping of hands by the feed rollers. Threshing cylinder has 2-4 chaff-cutter type blades and beaters. Chaff-cutter blades cut the crop into pieces and beater helps to detach grain from crop. All the threshed materials pass through the concave where it is subjected to aspiration action of blower. Light materials like chopped straw are blown away and grain etc fall on a set of reciprocating sieves. The clean grain is collected in trolley through auger elevator. It can be used to thresh the crop having high moisture content also. The machine is operated by PTO of a 35-hp tractor and is mounted on two pneumatic tyres for easy transportation. The grain output capacity is 16-20 q/h for wheat, 8-10 q/h for raya, 6-8 q/h for gram and 4-5 q/h for green gram. Threshing efficiency, un-threshed grain and visible damage are 98-99%, 1.5-2.0% and 1.0% respectively. Average total losses are about 5%. For moong crop, two cutting blades (out of four) are removed to reduce the damage. The average capacity of thresher is 4.0 q/h for moong and 5.0 q/h for guar crop. The breakage is less than 5.0% for moong and 0.5 to 2.0% for guar.

Fig. 8.50: High capacity 'Harambha' thresher
Courtesy: M/s. Amar Agricultural Implement Works, Ludhiana

Commercially available spike tooth type thresher has been used as a multi crop threshers after incorporating few modifications (Fig. 8.51). The diameter of threshing cylinder is 580 mm and length 326 mm. However, different sizes of threshers are available depending upon the crops to be threshed and power source available. The threshing cylinder has thirty-six spikes placed six in each row. For threshing pulses six spikes are retained on cylinder in 6 rows i.e. 1 in each row. These spikes are made of 16 mm studs. The arrangement of spikes on cylinder periphery is axial. Aspirator I & II are centrifugal type and having 4 & 3 blades respectively. Aspirator I separate the major chunk of chaff from sieve falling directly below the

Fig. 8.51: High capacity multicrop thresher in operation.
: Bharat Industrial Corporation, Moga

concave. Aspirator II accomplishes the stage II cleaning. It picks chaff from screen just prior to discharge of grain from main outlet. Reciprocating sieve system consists of replaceable set of sieve and screen mounted on oscillating frame. The amplitude of oscillations could also be varied. Upper sieve separates chaff bigger than grain while lower screen let the dust, smaller particles than grain pass through thus help in delivering clean grain at main outlet. The thresher is provided with four wheels made of cast iron for transportation and motor stand to fit the motor on it. It also has provision for attaching universal shaft for operating the thresher by tractor PTO. Threshing efficiency, un-threshed grain and visible damage are 98-99%, 1.5-2.0% and 1.0%, respectively. Average total losses are about 5%.

Axial flow sunflower thresher

The machine is used for threshing sunflower (Singh and Pandey, 2008; Garg and Singh, 2002). It consists of a feed hopper, bar type cylinder, thrower, two sieves, concave and a blower. It works on axial flow principle (Fig. 8.52). The cylinder of length 1500 mm has two portions, the first one of 1300 mm for threshing and the second of 200 mm for straw throwing. The threshing portion has raised spikes. The cylinder-concave clearance is 40 mm and is uniform throughout its length. The cylinder is of hexagonal shape and is fitted with seven louvers at a spacing of 180 mm. The louvers are made of 3 mm thick MS sheet and have a depth of 70 mm. The cleaning system consists of a centrifugal blower and three sieves inclined at an angle of 7-15°. It saves 70-80% of labour and operating time and 40-50% in cost of operation. The axial flow thresher gives 3-4 times more output compared to conventional thresher. The machine is operated by 7.5 hp motor/tractor and has a capacity of 600-900 kg/h of clean

Fig. 8.52: Axial flow sunflower thresher in operation.

grain. Recommended cylinder and blower speeds are 300-350 rpm and 1200-1400 rpm respectively. Threshing efficiency of the machine is 99% where as cleaning efficiency is about 90%. Grain losses vary from 0.65 to 2.94% depending upon crop moisture content. Machine requires 4-5 persons for crop feeding. It saves about 70% labour in comparison to traditional threshing.

Axial-flow high capacity paddy thresher

Machine is suitable for threshing paddy and it works on the principle of axial flow (Garg and Singh, 2002). It consists of a threshing cylinder, concave, cylinder casing, cleaning system and feeding chute (Fig. 8.53). In axial flow concept, the crop is fed from one end and the straw is taken out from the other end after completing threshing of crop. During this period the crop is rotated three and half times and all the grains are separated. The threshing cylinder is of peg type and has a diameter of 77 cm and length 150 cm. The first part of cylinder of length 133 cm has spikes for crop threshing and the 2nd portion of cylinder of length 17 cm has straw throwing blades. The concave is made of 8 mm round bars and cylinder-concave clearance is 40 mm. The casing of the thresher is provided with 7 louvers for moving the crop axially. For cleaning purposes, it is provided with 2 aspirator blowers and 2 sieves. The opening of the top sieve is 9 mm and of lower sieve 1.5 mm. Machine is operated by 35 hp tractor and can thresh about 13.0 q/h of clean grains. Recommended blower speeds are 1800-2000 rpm. Machine has a threshing efficiency of 100% and cleaning efficiency 99%. Total grain losses are less than 2.5%. It saves about 70% of labour and 20% of cost of operation and results 2-3% increase in grain recovery through reduced losses as compared to conventional method of manual threshing by beating.

Fig. 8.53: Axial-flow high capacity paddy thresher
Courtesy: Madho Agro Industries, Moga

Axial flow groundnut thresher

It is suitable for threshing of groundnut crop (Garg and Singh, 2002). It consists of an axial flow-threshing cylinder, concave, cleaning system, a blower and feeding platform (Fig. 8.54). The cylinder is of closed type and has pegs in the 1st portion of 130-cm length. The 2nd portion of 22 cm length has 5 throwers.

The cylinder concave opening is 3.0 cm. The casing is of hexagonal shape with 7 louvers inside for axial movement of the crop materials. It has 2 sieves. The upper sieve has 15 x 45 mm holes where as the lower sieve is made of 8-mm MS round bar having 15 mm spacing. The machine is operated by a tractor of 25 hp through a flat belt and pulley. The harvested crop at moisture content varying from low to high is fed to the threshing cylinder through feeding platform. The pods after threshing fall on the upper sieve through concave and straw is thrown out from the outer end. The output of the machine ranges from 170-220 kg/h with total pod losses of 1.0 to 2.25 percent. The pod damage is almost negligible. Threshing efficiency is 99-99.5% and cleaning efficiency 92-98%. Labour requirements with the machine varies from 32-42 man h/ha and it saves 70% of labour and 40% of cost of operation in comparison to traditional system.

Fig. 8.54: Axial flow groundnut thresher

Groundnut stripper (comb type)

It consists of a square frame with wire-loops on top edges of four arms (Anonymous, 2009). Stripping is done by holding the plants and pulling the pod portion over the loops manually (Fig. 8.55). Four persons can work at a time. It is suitable for detaching or stripping the pods from semi-dry groundnut vines. It is powered with 2 hp electric motor. It saves 66 per cent labour, 80 per cent operating time and 30 per cent on cost of operation and it also results in 10 per cent reduction in losses compared to conventional method of manual stripping.

Fig. 8.55: Groundnut stripper (comb type)

Groundnut-cum-castor decorticator

Fig. 8.56: Caster and groundnut decorticator

It is a manually operated oscillating type device having cast iron shoes with triangular projections for decortications of groundnut and castor pods to separate kernels (Anonymous, 2009). Separate concaves are provided for decorticating groundnut and castor (Fig. 8.56). It is not provided with cleaning device. It saves 98 per cent labour and operating time and 89 per cent on cost of operation compared to conventional method of hand shelling.

Caster sheller

Castor oil is an important export product used by pharmaceutical industry and cosmetic industry. Traditionally this crop is being shelled manually after drying in sunlight which is very labour intensive and low output method and need winnowing after shelling. Therefore, a manually operated caster sheller has been developed (Anonymous, 2009). It consists of feeding hopper, rubber coated disc type shelling unit and a blower (Fig. 8.57). Castor is fed to the shelling unit through feed hopper. It is suitable for shelling and winnowing of dried castor pods. The unit can be either operated manually or with a 0.5 hp electric motor. It saves 88 per cent labour and operating time and 69 per cent on cost of operation compared to conventional method of manually beating or rubbing with wooden plank. A commercial castor sheller (Fig. 8.58) is also available (Anonymous. 2010). It consists of feeding hopper, beater shelling

Fig. 8.57: Caster sheller

Fig. 8.58: Operation of castor sheller.
Courtesy: M/S Ganesh Raj Industries, Mehsana (Gujarat)

unit, concave, sieve, blower and PTO coupling unit. Three people are required to operate this machine. The threshing efficiency is 85.2% and cleaning efficiency 97.6%. The broken grains are in the range of 6-7%. The output capacity is 10-12 q/h.

Multi crop thresher for seed spices

The traditional practice of threshing of seed spices is either by beating on drums or wooden logs or treading by tractor wheels on mud floors. This method requires lots of labour and lot of dust and stone comes in during the threshing process which are removed manually by hand sieves or cleaning machine in factories reducing the quality of seed spices thus necessitates the mechanisation of seed spices threshing. Therefore, a tractor/electric motor operated thresher has been developed (Anonymous, 2010). The threshing cylinder has flat steel spikes with size of 30 x 120 x 6 mm (Fig. 8.59). The throat opening of thresher is kept as 580 x 240 mm. The diameter of threshing cylinder is kept as 660 mm and length as 550 mm. The thresher is mounted on rigid frame with covered power transmission system for threshing cylinder, three blowers and sieve shaking mechanism. The rpm of threshing cylinder is 540, blower 480 and drive pulley 440. The drive to power transmission is given by telescoping shaft connected to PTO of tractor and also by electric motor of 7.5 hp mounted on frame so wherever electricity is available, thresher can be used with electric power. It gives output of 240-260 kg/h of seed spices. The threshing efficiency is 100% and cleaning efficiency 99%. Blown out grain is nil. It saves 55%in cost of threshing and 72% in threshing time over manual threshing.

Fig. 8.59: Multi crop thresher for seed spices
Courtesy: M/S Makewell Industries, Unjha, Gujarat

Performance of threshing system

The performance of a threshing unit is the percent of seed detached from the non-grain parts of the plant and the percent of seed damaged. Two other important parameters are performance of cleaning and separating units. Power requirement is not a functional performance parameter. The performance of threshing system as defined by Bureau of Indian Standards in IS: 6284-1971 is given below:

Total grain input: It is the feed rate multiplied by grain content as obtained from straw-grain ratio.

$$\textbf{Unthreshed grain (\%)} = \frac{\text{Quantity of unthreshed grain obtained from all outlets (kg)}}{\text{Total grain input (kg)}} \times 100$$

Threshing Efficiency: It can be defined as the percentage of grain threshed, collected from all the outlets of a thresher with respect to grain input. It depends upon cylinder speed, concave clearance, spike shape, moisture content of crop, crop mass velocity, crop mat thickness etc. It can be calculated from un-threshed grain.

$$\text{Threshing efficiency (\%)} = \frac{\text{Total grain input (kg) - unthreshed grain from all outlets (kg)}}{\text{Total grain input (kg)}} \times 100$$

$$\text{Blown grain (\%)} = \frac{\text{Quantity of threshed grain obtained at 'Bhusa' outlet (kg)}}{\text{Total grain input (kg)}} \times 100$$

$$\text{Damaged grain (\%)} = \frac{\text{Quantity of damaged grain from all outlets (kg)}}{\text{Total grain input (kg)}} \times 100$$

$$\textbf{Sieve loss (\%)} = \frac{\text{Healthy grain obtained at sieve overflow + sieve underflow + struck grain (kg)}}{\text{Total grain input (kg)}} \times 100$$

Cleaning Efficiency: It is the percentage clean grain in the total grain obtained from the main grain outlet. It is affected by various factors such as crop mat thickness, crop mat velocity, cylinder speed, concave clearance, spike shape etc.

$$\text{Cleaning efficiency (\%)} = \frac{\text{Total grain received at main grain outlet (kg) - refraction At main grain outlets (kg)}}{\text{Total grain received at main outlets (kg)}} \times 100$$

Total Loss = Unthreshed grain + blown grain + cracked grain + sieve loss

Vegetable seed extractor and thresher

The extraction of seeds from different vegetables is commonly done manually, which is time and labour consuming and unhygienic. It often leads to injuries to the hand and seed productivity is low. An axial flow vegetable seed-extracting machine has been developed at Punjab Agricultural University, Ludhiana (Verma and Singh, 1988). The machine is quite simple and is capable of extracting seeds from a large number of vegetables such as tomato, brinjal, chillies, cucumber, summer squash, watermelon etc. It is based on wet extraction principle. The machine consists of feeding chute, primary cutting chamber, crushing chamber, seed collecting chamber, seed and water outlet (Fig. 8.60). The vegetable fruits are cut and crushed by means of axially arranged blades mounted on a rotor shaft. The shaft is rotated at a speed of 250-300 rpm. Conveying racks have been provided to move the pulp and coarse material along the axis. The waste materially is finally ejected from the waste outlet. Concave screens have been provided under the rotor for separating fine, medium and large seeds. Three 25 mm diameter pipes with holes, two on sides and one on top have been provided along the length of the rotor. Water is supplied to these pipes by a small centrifugal pump to wash seeds and finally crushed materials and collected separately. The machine can extract 1.25-3.00 kg of different vegetable seeds per man-hour.

The vegetable thresher (Fig. 8.61) design involves the principle of cutting, impact, separation and cleaning of vegetable crops to obtain seed (Anonymous, 1987-89). The design feature is such that it can handle dry crops of vegetables. Conventional wheat threshers employing chaff cutter blades can be suitably modified to thresh vegetable crops for seeds.

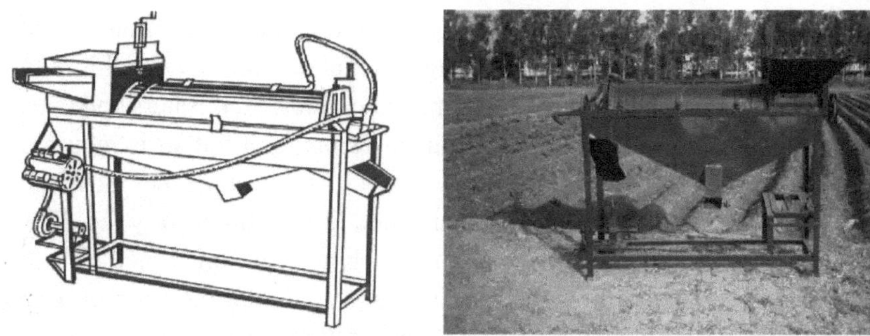

Fig. 8.60: Vegetable seed extracting machine.
(*Courtesy*: M/s National Agro Industries, Ludhiana)

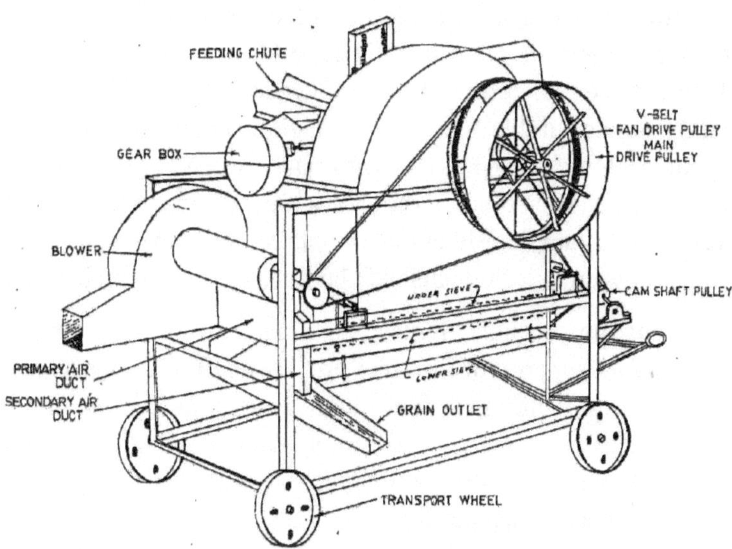

Fig. 8.61: Vegetable seed thresher.

Forage harvesting equipment

A variety of fodder crops like barseem, corn, sorghum, bajra are grown throughout the year on most of the Indian farms. Fodder harvesting practices followed in India are far different from that accepted in developed countries. Regular harvesting of fodder with traditional or improved sickles is continued throughout the year, which requires labour of the range of 160-200 man-h/ha. In advanced countries where dairy is well-developed most of the operations are fully mechanized. The machines used include self-propelled and/or pulled type windrower, shredders, hay conditioners, field choppers, harvesters and choppers, balers (along with bales handling machinery) and even machines to make wafers, raking machines are useful machines and so on. Balers and cube making machines are useful when low moisture storage of fodder is required or when transporting distances are large. These machines are not of much relevance to our conditions where fresh harvested fodders are locally fed. This limits our interest to cutting and chopping machinery. Also average size of land holding and other socio-economic factors is also a hurdle for the large size bailing and handling machines. Fodder harvesting includes operations of cutting the standing crop; its conditioning, transport and chopping. Therefore, in India simple and efficient forage harvesters are required that can harvest the crop economically and without causing much loss to the crop.

Shear cutting: To cut a material, forces are applied on it in shear at certain plane. Failing of the material in shear is desirable but generally shearing takes place along with certain deformation and bending of stems. This deformation and bending is undesirable which is wastage of energy. A common way of applying cutting forces is to press the material between two shearing elements, which move against each other without or with very small clearance. A pair of scissors is a good example. Either one or both of the elements may be moving. The motion between two elements may be linear, rotary or reciprocating.

Impact cutting: A single element shearing is also possible if the surface holding the plant or inertia of the stem provides sufficient resistance against the cutting element. If cutting is done close to the ground, ground itself serves as one cutting element. A high-speed cutting blade does not give much time to the plant stem to bend or deform before it is sheared off. This type of cutting is commonly known as impact cutting.

Impact type cutters

Rotary cutters: Rotary cutters have blades rotating in a horizontal plane, installed on periphery of discs or drums rotating on their vertical axes. Rotary cutters used for forage harvesting usually have counter rotating pair of drums or two pairs of discs. Each drum or disc usually has two, sometimes 4 knives. However, machines having more sets of cutting drums may have width of cut up to 6 m. Peripheral speeds vary from 50 to 80 m/s depending upon type of crop, its maturity etc. Blades are sometimes hinged through vertical pins on periphery of discs to provide a back swing if some hard object is encountered. The counter rotating drums have a tendency to make a windrow in between.

Nice windrows can be obtained by using shields depending upon type of crop. The drums get their power through a common shaft rotated by PTO through V-belts. From common shaft drive is given through chains or bevel gears to drums so that relative rotating of drums is positively fixed. This avoids fouling of blades of two adjacent discs or drums, which have slight overlap. These types of cutters differ from conventional cutter bars in many ways. Unlike cutter bars they can be operated at high forward speeds (up to 14 km/h). There is hardly any chance of choking of the machine if proper speed of rotation is maintained. They need comparatively low maintenance. But their initial prices and power requirements are quite high in comparison to simple cutter bars.

Flail shredders/mowers

Flail type mowers have large number of blades staggered in 3-4 lines on a drum rotating on a horizontal axis perpendicular to the direction of motion. Width of blades varies 50 to 150 mm and they also very in shape. These blades or knives are attached to the drum through hinges or small ring links to avoid the damage if any hard object comes in contact. Effective width of such machines varies from 1.2 to 4 m. Initially flail shredders were used to cut weeds before plowing or mowing lawns. The peripheral speeds of the knives were 50-60 m/s. When such machines are modified for fodder; the peripheral speeds are reduced to 40 m/s to avoid multiple cutting and field losses of the crop, even though the field losses are more than the conventional cutter bars. Flail mowers also help in conditioning of the hay as they spread the crop in uniform windrow. Power requirement of these machines is also very high, about 25 kW per meter width. It depends upon many factors like crop density, forward speed etc. Their high initial costs are justified for their performance under difficult conditions.

Convention cutter bar mowers

It has a reciprocating type cutter bar, which is used to reap fodder as well as other crops. After harvesting it spreads the crop on the ground. Animal-drawn, tractor mounted as well as small engine operated cutter bars are available these days. Tractor operated machines are generally rear-side mounted or front mounted. But front mounted machines provide better visibility to the operator and have easy control but additional mountings and drives are required to operate. Rear mounted cutter bars are most common. Effective width of cutter bars varies from 1.52 to 2.75 m (5-9 feet). The offset drag force on the outer end of the cutter bar limits its width. Two shoes of the cutter bar help to maintain the uniform height of cut. During transport cutter bar can be folded to vertical position and when in operation it is spread to horizontal position.

Register and alignment: Register is the term, which relates the relative position of knife section from the ledger plates around it to its speed. During reciprocating motion speed of the cutter bar varies in sinusoidal style. A knife is said to be in proper register if it attains its maximum speed when exactly between two guards or ledger plates. The idea is to avoid cutting when speed of knife is low. For the adjustments entire cutter bar can be moved in or out with respect to the Pitman shaft. It is essential for even cutting. Alignment of the cutter bar is important to keep it exactly at right angle to the direction of motion while in operation. Since entire cutter bar is projected out from Pitman shaft side it has a tendency to bend backward under the crop pressure. To counter this deflection outer end of the cutter bar is kept a little ahead of the inner one when in stationary position. Outer end is kept 20 to 50 mm ahead per meter width of the cutter bar depending upon the forward speed and density of the crop.

Operating speeds and adjustments: Cutter bars are generally operated from the PTO through Pitman shaft. This is the cheapest mechanism but there are certain limitations. Pitman shaft and cutter bar should be ideally in line. Pitman drives can be operated at 850 to 1000 rpm. With dynamic balancing speed can be increased to 1200 rpm. A typical speed of 1000 rpm of Pitman at 76.2 mm stroke length gives maximum top speed to the knife about 4 m/s. Even this top speed of 4 m/s is not capable of cutting stems by impact as in rotary cutters (linear speed 50 m/s). Thus close clearance between knife sections and ledger plates needs to be maintained. Adjusting knife clips can reduce the clearance. Larger clearances result in clogging of knives and there can be considerable increase in power requirement due to deflection and deformation of stalks. The forward speed ratio to the strokes per unit time in PTO operated machines is adjustable. Thus cutter-bars can be operated at slow forward speed in dense crop and relatively high forward speeds if the crops are not that heavy. But deflection of stalks before cut and bunching of stalks in a particular stroke limits the top forward speeds which are in most of the cases below 6-7 km/h. Higher forward speed may cause frequent clogging of the machine. If crop is not adequately erect forward speed may further be reduced. Sharp and well-maintained knives with close adjustments help to maintain good forward speed.

Energy and power requirement

Power required for operating a filed chopper could be divided into the power requirements for gathering and conveying of crop, shearing/chopping of material, accelerating the cut-material, pumping air along with accelerated material, overcoming friction to move chopped material and overcoming internal mechanical friction within the machine. Actual power requirements depend upon the type and condition of crop, type and condition of machine, its adjustments and length of cut. Under same conditions, the power requirements will increase as the feed rate or machine output per unit time increases. For a typical field chopper, power requirements under test conditions (13 mm length of cut, 35 m/s peripheral speed, crop moisture 74%) are measured to be 1.12 kWh/ton. Out of the total energy requirements about 35% power is required for cutting, 50% for accelerating the cut material (including its friction) and blowing air and the remaining 15% is used in gathering and conveying the crop. As moisture content increases energy requirements increase because coefficient of friction increases rapidly above 30% of moisture content but increase is very small beyond 50% m.c. Also, shear strength of materials content beyond certain value may cause decrease in energy requirements per unit of dry matter. Energy requirements decrease as length of cut is increased. As length of cut increases, energy used in cutting (shearing) per unit mass decreases whereas other requirements like acceleration of material, pumping of air, feeding to the cutter head remain the same. Thus decrease in overall energy requirement is at decreasing rate as length of cut increases.

Energy and power requirements in reciprocating cutter bar are quite low in comparison to rotary cutters, especially shredders, which need a lot of power to give acceleration to cut material. Since power requirements and thus energy requirements per unit mass of cut crop may vary widely depending upon type and condition of crop, condition of cutterbar, forward speed etc, it is not easy to generalize the figures. Different research workers have also given contradictory results. But as a rough idea, on an average 0.91 kW per meter of cutter bar width may be required out of PTO at forward speed of 7 km/h and 100 rpm of the crankshaft. However, peak values (one peak per stroke) lie somewhere near 2.4 kW/m. Out of these loads two third loads was to cover frictional requirements of the moving parts and one third for cutting of the crop. Additional power will be required to pull the machine forward. These values are based on field tests. The actual power requirements for cutting in the laboratories is found to be even less than these values.

Tractor operated fodder harvester

Mechanization of fodder cultivation system is of utmost importance for successful white revolution. The major fodder crops are oats, barseem, maize, sorghum, bajra and ginni grass. The harvesting of fodder crops has become a serious problem because of labour shortage and high labour requirement (120-150 man-h/ha). Moreover, during summer, harvesting of tall fodder is more serious because of high humidity, scarcity of labour and higher wages problems. Except scythe (Fig. 8.62) no harvesting equipment is in use for barseem. It is much faster than sickle and is operated in the standing posture conveniently. Labour required for sickle and scythe is measured to be 106 man-h/ha and 57 man-h/ha respectively. The available mowers are not used because once the crop is harvested; it has to be cropped again after collection. Since the crop cut with mower does not fall in a windrow so labour requirement for collection becomes quite high. On the other hand mower-cum-chopper-loader performs all the three jobs i.e. to harvest, chop and filling trolley.

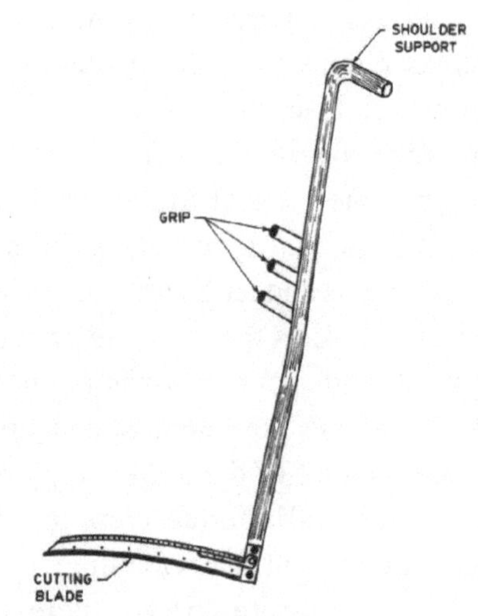

Fig. 8.62: Scythe

The flail type harvester-cum-chopper-cum-loader has been designed to facilitate harvesting and chopping and loading operations simultaneously (Anonymous, 2008, 2010). The machine consists of a rotary shaft mounted with blades (flails) to harvest the crop, auger for conveying the cut crop, cutters for chopping and conveying chopped fodder through outlet on to the trailer (Fig. 8.63). The blades on the rotary shaft are staggered in 3

Fig. 8.63: Tractor operated flail type fodder harvester in operation.

rows of 13 blades on each row on a horizontal axis perpendicular to the direction of motion. After the blades cut the crop comes to the auger that conveys it to the chopper unit. The chopping mechanism having 4 blades cuts the crops in to pieces and the chopped material is thrown out with high speed and is filled in to the hitched trailer to the machine. The size of cut fodder varies from 4.8 to 10.6 cm. The effective width of coverage varies 1.12-1.26 m with average fuel consumption of 4.5 l/h. The operation of machine for 22 min fills one trolley with chopped fodder that indicated saving of 3-4 times compared to traditional practice. It also harvests lodged and over matured crop without any difficulty. The capacity of the machine is 0.20 ha/h at a forward speed of 2.5-4.0 km/h. Any tractor of 45 hp and above can operate the machine.

Self-propelled fodder harvester (cutter bar type)

The self-propelled fodder harvester consists of 1450 mm cutter bar (Fig. 8.64). It is mounted in the centre of the power transmission with the help of two side linkages made of high-pressure circular pipes to dampen the vibration of cutter bar. The unit is also provided with crop gathering drum-having flappers rotating inwards near the cutter bar (Anonymous, 2008, 2010). The vertical side covers around the transmission system are provided to avoid blockage of the fodder and extended vertical guides are provided for proper windrowing. The ground clearance of the machine has been increased to 480 mm to facilitate easy passage of fodder. Power to wheel is provided through extension with the help of chains and sprockets with the chassis suitably strengthened. The track width is also widened to 1040 mm and a caster wheel is provided at the rear to improve stability and manoeuvrability. The effective field capacity of machine is 0.1 ha/h at a forward speed of 1.5-2.0 km/h.

Fig. 8.64: Self-propelled fodder harvester (cutter bar type)

Self-propelled lucerne harvester

Lucerne is perennial leguminous plant growing to a height of 60 to 90 cm. It is a forage crop generally grown in basins of size 1.20 x 6 m having a very high

Fig. 8.65: A self-propelled lucerne harvester in operation.

yielding potential under appropriate fertilizer and irrigation management (Anonymous, 2008, 2010). Harvesting of crop is one of the important agricultural operations, which demands considerable amount of labour because it is done manually by sickles. The cutting and laying in the windrows consume 65-75% of labour and gathering, bundle making and transport in the field involve the rest of the labour requirement. The scarcity and high cost of labour and drudgery during harvesting are the serious problem faced by the farmers. It is, therefore, essential to adopt the mechanical methods so that the timeliness in harvesting operation could be ensured and field losses are minimized to increase the productivity and production on the farm. The self-propelled Lucerne harvester consists of a gearbox and cutter bar (Fig. 8.65). Type of cutter bar is bi-directional reciprocating type made from high carbon steel. Length of stroke for cutter bar is 25 mm and effective width of cutter bar is 860 mm. A man can walk behind the machine with an average speed of 2 km/h. The recommended speed ratio of the average cutter bar speed to the forward speed of machine is 1.3: 1.4. Two wheels are used for transportation purpose. The ground wheels drive the reel of the harvester. Ground drive provides the desirable feature of maintaining a constant speed ratio between peripheral speed and forward speed. The reel is made up from Ø 6 mm MS bar of diameter 70 cm and length 74 cm. The speed ratio of the ground wheel to reel is 1:1. The effective field capacity is 0.113 ha/h and field efficiency 70-75%. There is net saving of 52% in cost and net time saving is 90%.

Self-propelled forage harvester for fodder crops (barseem, barley, sorghum & maize)

A self-propelled machine operated by 12 hp Lombaridine engine has been developed at Punjab Agricultural University, Ludhiana for harvesting fodder crops like maize, bajra and oat (Anonymous, 2008, 2010). It cuts the crop and scatters behind the cutter bar. The machine consists of a self-propelled forage harvester having a horizontal rotor on which nine L-shape flails are mounted in staggered position in three rows (Fig. 8.66). The maneuverability of the machine is controlled by a mechanism provided at foot of the operator. The cutting width of the machine is 60 cm. The crop is cut by the flails by impact and guided by the shield provided on the rotor. Harvesting efficiency and throughput capacity of the machine for harvesting maize, bajra and oat fodder increases with the increase in flail speed and it varies from 95 to 100 percent. With the increase in flail speeds, the height of cut from the ground decreases from 7.94 to 6.32 cm thus resulting in higher throughput capacity of the machine. There is a saving of about 48 percent in labour cost. The machine has a very low capacity and the collection of fodder difficult. An attachment has been developed to collect the fodder but resulted in increase in power requirement, as engine has to develop additional power for conveying of cut fodder.

Fig. 8.66: A self-propelled forage harvester during field operation.

Tractor operated cutter bar type forage harvester and chaffer-cum- loader

The system consists of a two separate units (Anonymous, 2010). First one consists of a mower (Fig. 8.67) and the second one a chaff cutter-cum-loader (Fig. 8.68). Both the machines are tractor operated. The mower has a working width of 196 cm and so designed that it can be folded easily during transportation. The chaffer-cum-loader has a chaffing cross section area 28 x 13.5 cm^2. It has two powered feed rollers and one compressing roller. It has two chaffing blades for chaffing of fodder and six thrower attachments are being mounted on the periphery of the cutter. A reversing mechanism has also provided for safety. The fodder is harvested with mover and collected in the field itself. The chaffer-cum-loader chaffs the fodder at site and the throwers mounted on the periphery of the cutter throw the fodder directly into the trailer. The field capacity of the side mover varies from 0.20-0.25 ha/h and fuel consumption from 3.5-4.0 l/h when operated within the speed range of 1.5 to

2.0 km/h. The throughput capacity of the chaffer-cum-loader varies from 12.0-18.0 t/h. The crop after chaffing is directly loaded in the trailer. The chaffer cut the fodder in pieces of size of 18-20 mm. There is a saving of about 49 percent in cost of operation and 65 percent in labour cost. The machine is used for harvesting and chopping of fodder crops like maize, jowar and bajra.

Fig. 8.67: A view of side mounted mower harvesting oat crop

Fig. 8.68: A view of chaffer-cum-loader in operation

Tree climber

In order to harvest coconut and arecanut at faster rate with proper safety, a tree climber has been developed. The developed tree climber is free from any accident risk during its operation (Anonymous, 2010). The climber made of M.S. square pipe consists of two components (Fig. 8.69). Adjustable belts connect the components. The upper component is provided with a seating arrangement and lower component is having provision for holding the foot. The rubber cushioning is provided at the portion of frames, which comes in contact with tree to avoid any damage of tree. By standing on the lower component, the upper component can be moved up or down over the tree. The operator can safely climb a tree of 10 m height in 1.5 min without any risk.

Fig. 8.69: Tree climber in operation.

Manual climbing device for Palmyra

Harvesting of palmyra fruits is done by trained labour who climb the 60 foot tree, which is an accident prone operation demanding a strenuous effort. Moreover skilled tree climbers are rarer nowadays because of the drudgery involved in this operation. A coconut tree climbing device has been developed successfully at TNAU Coimbatore, which is a simple device facilitating any untrained person to safely climb the coconut trees and do the required operations. However unlike coconut trees, the variation in the girth of palmyra tree from base to top is considerable and hence the developed device has been re-designed so as to adjust the frame width on-the-go to correspond with the girth of the tree at each height. To accommodate a wider girth range, the mechanism has been provided with two number of sliding inclined bushes, which provide V shape to grip on the tree. Simple cranks and flexible shafts provided for easy adjustment (Fig. 8.70). The average time taken for climbing up and down is about 6.30 min for a 12 m tree with time for fixing and removing the device on the tree being 4 min.

Fig. 8.70: Palmyra tree climber in operation

Aerial access hoist for coconut and tall tree crop management

All existing tools and devices involve the operator to climb up the tree for harvesting and carrying out other management practices at the crown. Farmers who own large areas are interested in having a system which can elevate a person up to the tree crown by a portable aerial access platform. The existing aerial access platforms are having the limitations i.e. the access is vertically upward and not in the side wards. The machines are designed to operate by resting on firm surface, most machines have very wide stabilizing legs which cannot be operated under field conditions and require long time for setting up and operating. Keeping the above constraints in view and after a thorough study of the planting pattern and space requirements technical requirements of a tractor mounted aerial access hoist were formulated. The design of individual sub systems and structures has been done by M/s Vanjax, Chennai in collaboration with TNAU Coimbatore (Anonymous, 2010). A full length chassis has been extended from the front of the tractor to the rear and bolted to the tractor chassis (Fig. 8.71). This forms the support for the hoist and provided for mounting of all stabilizers. The entire weight of the hoist and moments are transmitted through the chassis to the stabilizers without transferring to the tractor chassis. The tractor mounted aerial access hoist can be taken into the coconut field and four trees can be accessed from a single position. The time required for locating unit and operating stabilizers is 1 min. The time required for positioning against a tree of 10 m height is 2 min. Suitable safety devices i.e. hydraulically operated 4 nos. stabilizers have been incorporated to ensure stability of the hoist. The positioning of the operator platform can be done by the operator himself using electro hydraulic controls.

Fig. 8.71: Tractor operated aerial access hoist for coconut and tall tree crop management

Tractor operated telescopic hoist

The tractor operated telescopic hoist can be used in orchards and plantation crops for trimming, pruining, spraying and harvesting of fruits (Bhardwaj *et al.*, 2004). It consists of two square aluminium ladders each made of U-section as frames and round hollow pipes as cross members (Fig. 8.72). The U-sections of the outer ladder are inward facing while U-sections of the inner ladder face outward, sliding one over other. Two wire ropes are provided that are driven with a hydraulic motor. The motor while running in clock-wise direction helps in lifting the inner ladder and while running

Fig. 8.72: Tractor operated telescopic hoist

in anti-clockwise direction lowers the inner ladder down. A support frame fitted to the platform on the top end of the inner ladder helps as a safety frame. It can cover up to 9.5 m maximum height and minimum height of platform is 5.5 m.

Self-propelled platform type fruit harvesting device

Harvesting, pruning, spraying and other canopy management in fruit trees such as mango, citrus and sapota are difficult and labour intensive. These operations contribute major cost in fruit production system. Available mechanical harvesting devices have limited access due to tree height and canopy. Hand picking continues to be the only solution of fruit harvesting in the country. A self-propelled platform type hydraulically operated multipurpose machine has been developed for harvesting, spraying, pruning and canopy management for medium height fruit trees (Anonymous, 2013). It has a vertical reach of 6 m, loads carrying capacity 200 kg and could be operated at ground speed of 3 km/h (Fig. 8.73). It is hydraulically powered by 8.7 kW petrol engine and has low centre of gravity (29 cm) for better stability during operation in orchard. Lifting & lowering of platform, forward & backward movement of machine and steering are controlled by operator himself from the platform. The machine is easy to operate and requires low maintenance. The operator can pick 700-1100 mangoes/h depending upon the fruit density on the tree.

Fig. 8.73: Self-propelled platform type fruit harvesting device

Tractor mounted banana stem shredder

Banana is a major cash crop of the country. In India about 20 cultivars viz., Dwarf Carvendish, Rabusta, Monthan, Proven, Nandran, Red banana, Nyali, Safed Vekhi, Basarao, Ardhapuri etc are cultivated. Banana is mostly check row planted with the spacing varying from 125 x 125 to 150 x 150 cm, depending upon the variety. Plant population is about 4,500 per ha. After the harvest of the banana bunch, the stem is manually cut and left in the rows. After the harvest of the whole field, these are collected and left near the boundary for drying and subsequent burning. This process is tedious and time consuming. The banana stem shredder helps in disposing of the stem immediately after harvest. Shredded material is suitable for mulching in the banana garden and also for vermi compost. The average diameter of banana stem is 225 mm at the bottom and 100 mm at the top with the average height of 240 cm. The banana stem consists of 95% of water and only 5% of fiber. A tractor operated banana stem shredder has been developed (Anonymous, 2006). The shredding unit consisted of 4 blades placed perpendicular to each other at 225 cm distance (Fig. 8.74). Additionally, 12 numbers of spikes with flat cutting edge are fitted with a gap of 120 mm between the rows. The whole device is mounted on a frame made of MS angle. The blades are driven by the PTO of the tractor with a bevel gear box and the hopper is trapezoidal in shape with a height of 800 mm. The stem is cut into small pieces and the water and fibre are separated. It takes 1.2 minutes to shred the stem having average height of 2400 mm. About 52 stems are required for shredding in one hour. The shredded material can be used for mulching in the banana garden. The shredded material takes 3-4 days to dry. The shredded fibre can be used for preparing vermi compost.

Fig. 8.74: Tractor mounted banana Stem shredder

Tractor operated banana clump remover

Banana crop is maintained for two years to get the benefit of two harvests. The crop needs removal of clumps (plants along with root portion) after two years. During the process of removal of the clump, the entire mother plant along with the rhyzome and side suckers as a whole mass has to be removed so as to prepare the land for the next crop. Manual labourers using crowbars and spades do this operation. The labourers have to dig to a depth of up to 450 mm to remove the clumps and hence the operation is cumbersome. In view of above, a tractor operated banana clump remover has been developed at TNAU, Coimbatore to mechanize the clump removing operation and to reduce human drudgery involved (Anonymous, 2006). The nine-tyne cultivator frame has been adopted for the development of the equipment. Two numbers of 100 x 15 x 1000 mm sub-soiler shanks with shares of size 190 x 40 x 5 mm are fitted in the nine tyne cultivator frame at 225 mm spacing (Fig. 8.75). These two sub soilers perform as a fork while removing the banana clump. A deflector has been provided to push the soil sideways. The equipment is attached to the 3-point linkage of a 26 kW tractor. For removal of banana clump the sub soiler shank is positioned behind it and pressed into soil with the hydraulic system and the tractor is gently moved forward, simultaneously lifting the sub-soiler. This combined action removes the entire clump along with its root portion as a whole mass. The field capacity of machine is 0.5 ha/h.

Fig. 8.75: Tractor mounted banana clump remover in operation.

Dehusker for fresh arecanut

Most of the arecanut produced is dehusked in green state. Labour requirement for dehusking is estimated at 7-8 kg of nuts per day per labour. This involves huge labour requirement and high cost (Anonymous, 2010). Existing models are expensive and also cause damage to the kernels. Two different rotary dehusking mechanisms have been developed. The model-I has circumferential serrated blade and the model II has longitudinal profiled blades. The concave has two spring loaded rubber pads that press the fruit gently against the rotor (Fig. 8.76). The fruits are fed at the top manually. The fruits travel half the circumference and

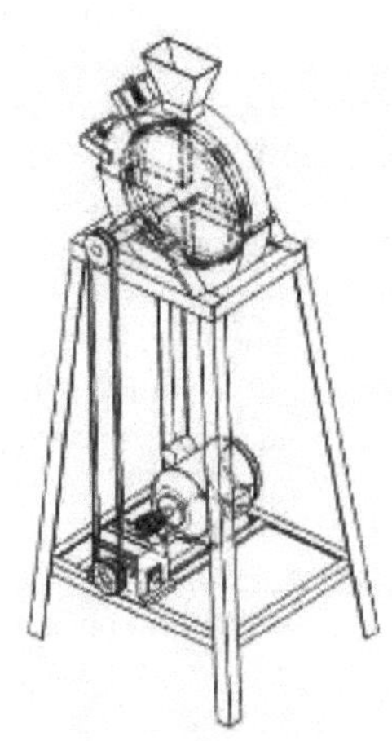

Fig. 8.76: Dehusker for fresh arecanut

are dehusked in the process. The dehusked kernals fall to the bottom along with the husk. The dehusking efficiency is 39 to 46% in Model-I and 53 to 67% in model-II. The best dehusking performance is observed at 60 rpm. The breakage is in the range of 8 to 10%.

References

Anonymous. 1987-89. Biennial Report. Department of farm Power & Machinery, Punjab Agricultural University, Ludhiana.

Anonymous. 2006. Research Highlight. AICRP on Farm Implements and Machinery, CIAE Bhopal. Technical Bulletin No.: CIAE/FIM/2006.

Anonymous. 2008. Research Highlight. AICRP on Farm Implements and Machinery, CIAE Bhopal. Technical Bulletin No.: CIAE/FIM/2008/141.

Anonymous. 2009. Technical Folder. AICRP on Farm Implements and Machinery, CIAE Bhopal. Technical Bulletin No.: CIAE/FIM/2009/144.

Anonymous. 2010. Research Highlight. AICRP on Farm Implements and Machinery, CIAE Bhopal. Technical Bulletin No.: CIAE/FIM/2010/151.

Anonymous. 2012. Directory of Successful Farm Machinery in SAARC Countries. SAARC Agriculture Centre. BARC Complex, Farmgate, Dhaka – 1215 (Bangladesh).

Anonymous. 2013. Research Highlight. AICRP on Farm Implements and Machinery, CIAE Bhopal. Technical Bulletin No.: CIAE/FIM/2013/158.

Banga K L; Mittal V K; Sharma V K. 1984. Study of selected parameters affecting performance of spike tooth type wheat threshing system. *J. Agric. Engg*. 21 (1&2): 25-43.

Garg I K; Singh Surendra. 2002. Farm equipment for Punjab agriculture. Department of Farm Power & Machinery, Punjab Agricultural University, Ludhiana.

Gupta P K; Singh Surendra; Sharma V K. 1985. Performance studies on tractor operated combine for maize threshing. Agril. Engg. Today (ISAE). 9(4): 40-42.

Manes G S; Bal A S; Sandhar N S. 1995. Feasibility of using spike tooth thresher for different crops. Paper presented at XXXI Annual Convention of ISAE, held at K.A.U. Thrissur, Kerala. Dec. 28-30.

Mehta M L; Sharma V K. 1985. Studies on delinking of cutting and threshing system of chaff cutter type wheat thresher. *J. Agic. Engg*. 22(2):10-18.

Mehta M L; Sharma V.K. 1986. Studies on threshing system of chaff-cutter type wheat thresher. *Journal of Research, Punjab Agric. Univ*., 22(4) : 735-741

Mudgal V D; Jain N K; Bordia J S; Seth P. 1998. Research Digest: AICRP on Post Harvest Technology (1992-97) Udaipur Centre. CTAE, MPUAT, Udaipur, 26-32p.

Pandey M M; Majumdar K L; Singh Gyanendra; Singh Gajendra. 1997. Farm Machinery Research Digest. Technical Bulletin No. CIAE/97/69, Central Institute of Agricultural Engineering, Bhopal, 328 p.

Singh C P; Garg I K. 1977. Design, development and field evaluation of simple bullock-drawn and tractor-drawn potato diggers. *Ind. J. of agric. Sci*. 47(2): 104-08.

Singh P R. 2014. Souvenir, Tractor and Agricultural Machinery Manufacturers' Meet (TAMM 2014). Held at IISR Lucknow during February 8-9.

Singh Surendra. 2007. Farm Machinery – Principles and Applications. Directorate of Information & Publication of Agriculture, Indian Council of Agricultural Research, Krishi Anusandhan Bhawan-I, Pusa Campus, New Delhi.

Singh Surendra; Pandey M M. 2008. X Plan Achievements (2002-2007). AICRP on Farm Implements and Machinery, CIAE Bhopal. Technical Bulletin No.: CIAE/2008/137.

Singh Surendra; Verma S R. 2009. Farm Machinery Maintenance & Management. Directorate of Information & Publication of Agriculture, Indian Council of Agricultural Research, Krishi Anusandhan Bhawan-I, Pusa Campus, New Delhi.

Singh S P. 2010. Ergonomical interventions in developing hand operated maize dehusker-sheller for farm women. Unpublished Ph. D. thesis. MPUAT Udaipur.

Verma S R; Garg, R L. 1970a. Tractor operated groundnut digger shaker. *Indian farming*, 11(11):33-35.

Verma S R; Garg R L. 1970b. Development and performance of tractor mounted groundnut digger shaker. *J. Agric. Engg. ISAE*, 7(2): 32-34.

Verma S R. 1979. Design, development and performance of potato harvesting equipment. Proceedings of International symposium on post-harvest technology and utilization of potato. International Potato Centre, Region VI New Delhi and CPRI, Shimal, Aug. 30-Sept. 2: 117-134.

Verma S R; Datta R K; Gupta C P. 1977. Performance of an experimental potato digger with oscillating blade. *J Agric. Engg.* XIV (3): 99-107.

Verma S R; Rawal G S; Bhatia B S. 1978. A study on human accidents in wheat threshers. *J. Agric. Engng. ISAE* 15(1): 19-23.

Verma S R; Singh Hari. 1988. Development of an axial-flow vegetable seed extracting machine. *J. Agric. Engng. ISAE* 25(1): 98-104.

Yadav R N S. 2003. Sugarcane Production Machinery, CIAE, Bhopal. Bulletin No. CIAE/NATP/-SM/2002/95.

9 Straw Management Equipment

Management of paddy straw is one of the important problems being faced by the farmers. Burning of rice straw in the field is the traditional method being practiced by most of the farmers, which causes atmospheric pollution and nutrient loss. There are three options available to manage the straw viz. i) burning the straw in the field, ii) collecting/baling the straw from the field, and incorporation of straw in the field. If rice straw is not burnt then baling may provide an attractive economic environmentally safe option. Manual collection of paddy straw is a laborious job and also storing the large volume of straw is another problem. As for as paddy straw is concerned, it has higher silica content so animals do not prefer it. Even then there are wide usages of the straw in paper mills for cardboard manufacturing, for packaging the materials, for mushroom cultivation, for burning in boilers, for animal feeds in drought regions etc. So baling the straw and compacting in small, transportable size and shapes is required. Straw can be collected in the field using machines viz. flail type paddy straw chopper, stubble shaver and straw baler (Anonymous, 1999; Singh 2007). The average length of loose straw is 55 cm and height of standing stubbles 27 cm. The weight distribution of paddy straw after combine harvesting shows that the 44% is loose straw and 56% is contributed by standing stubbles. The total yield of paddy straw is about 125 q/ha. The yield of loose straw is 55 q/ha and standing stubbles is about 70 q/ha.

Tractor operated stubble shaver

It is used for harvesting paddy straw after combine harvesting and is operated by a 35 hp tractor. It is a tractor rear mounted PTO operated machine. The machine consists of 2-3 blades mounted on the ends of a rotating arm hinged with pivot pins (Fig. 9.1). The power to the blades is supplied from PTO of a tractor through a speed reduction gearbox and belt and pulley drive giving a speed ratio of 1.8:1 (Anonymous, 1987-89; Pandey *et al.*, 1997). The blades rotate in a horizontal plane near the ground and cut or save the stubble left after combine operation in paddy fields. The pulley and tip diameter of blades is 250 and 1500 mm respectively. The tip speed of blade is 76 m/s at PTO rated speed. The overall assembly is covered with an MS sheet to protect the machine from dust and stubble. The height-adjusting wheel has been provided to adjust the height of cut. The machine has been provided with fenders in front and two sides as safety measures for diverting the trash on the rear. It can cover about 0.2-0.3 ha/h. The blades need sharpening after every 10-12 ha of use and need replacement after 100 ha. The use of stubble shaver saves about two disking operations, which saves time, energy and money.

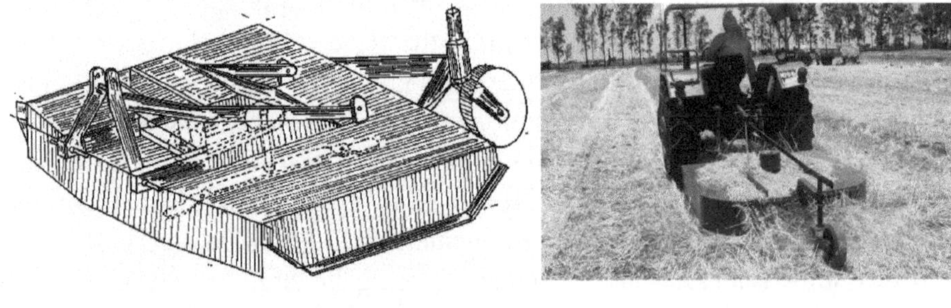

a) Isometric view of machine b) Machine working in field

Fig. 9.1: Tractor operated stubble shaver

Tractor operated paddy straw chopper-cum-spreader

Rice is mostly combine harvested in northern India and the straw left in the field is a major obstacle for subsequent farm operations. Incorporation of straw in the field requires 6-7 operations which involves high cost and is highly time consuming. Hence farmers resort to burning of the straw which is not an environmentally benign operation and results in the loss of organic matter. The commonly used equipment (offset disc harrow, rotavator) is unable to work in straw laden field conditions. PAU, Ludhiana developed a tractor operated straw chopper-cum-spreading machine in association with a manufacturer (M/s Dashmesh Mechanical Works, Amargadh) for effective straw management in the field (Singh, 2002, 2003, 2004). In a single operation, the stubbles left after combining are chopped in to tiny pieces and spread on the field. The machine consists of a rotary shaft mounted with blades known as flails to cut the straw and knives for chopping (Fig. 9.2). The machine being operated with a 33 kW tractor chops the stubbles at a flail speed of 1000 rpm and chopper speed of 2000 rpm. The cut straw passes on to the chopping mechanism. The cutting unit has 38 flails mounted in 3-rows. The chopping mechanism has 660 mm diameter cylinder with 6-rows of serrated knives and 4 counter rows each having 22 knives fixed at the bottom. Each row on the cylinder has 23 knives and knives are spaced at 90 mm distance. The working width is 2.28 m. The straw is chopped to a size of 70-100 mm and spread in the field. The field capacity is 0.4-0.6 ha/h at about 2.5 km/h speed of operation.

Fig. 9.2: Tractor operated straw chopper

Mulch seeder for sowing of wheat

In North India maximum soil temperature at 5 cm depth during summer months exceeds 45 degree C. Straw mulch reduces the amount of radiation reaching the soil surface and, therefore, reduces the maximum soil temperature. On a clear sunrise day temperature reduction can be more than 10 degree C at 5 cm depth (Anonymous, 2009). The beneficial effect of residue mulch on soil moisture and temperature changes affect different plant processes like seed germination, seedling emergence, root growth, N fixation by the microbes, etc which ultimately determine growth and yield of crops. It has been observed that symbiotic parameters of biological nitrogen fixation can be improved with mulching and are responsible for better yield of pigeon pea. Considerable research information is available about the effect of organic matter (FYM compost) on soil biological, chemical and physical properties as well as crop yields. The role of organics in increasing soil productivity is proved beyond doubt. Major amount of rice straw and of wheat straw is being burnt in northern India which not only pollutes the atmosphere but also result in nutrient loss.

A new machine has been designed and developed (Fig. 9.3) for sowing of wheat crop in combine harvested paddy fields (Singh, 2003 and 2004). This is called mulch seeder which chops the paddy straw and throws it on the sown seedbed, which has seeding equipment underneath. This is a combination of flail type chopper, conveying chute and underneath a seed drill/bed planter attachment. This is operated by 50-60 hp tractors. The width of the flail chopper is 1.85 m. Total 32 flails are mounted on a drum in 4 rows in staggered manner. The rotational speed of flails shaft is more than the double of PTO shaft of the tractor (about 1100 rpm). These flails after chopping the paddy straw (loose and stubbles) feed to the conveying long chute. The long curved chute conveys the chopped paddy straw to the ground. Under this chute a 3-point link system is provided for attachment of any type of sowing implement such as seed-cum-fertilizer drill, no-till drill or bed planter. The raising and lowering of this implement is controlled by auxiliary hydraulic connection provided from the hydraulic system of the tractor. The straw chopped by flails conveyed to rear end of the implement. This loose straw on sown seedbed provides mulch. This mulch provides a cover to soil, which maintains its temperature and reduces the moisture loss from the soil surface. Due to availability of mulch the weed population is also reduced. This machine without seeding attachment can also be used for collecting the straw in the trolley (Fig. 9.4) instead of mulch and later can be used as an animal feed or for any other purpose. Its field capacity is 0.25 to 0.35 ha/h.

Fig. 9.3: Tractor operated mulch seeder

Fig. 9.4: Tractor operated straw collector and loader

Paddy straw chopper-cum-loader

Combines are the quick means of harvesting the rice crop but the straw left go as waste, which otherwise could be utilized as cattle feed. Therefore, efforts need to be made to develop/identify equipment, which can recover maximum quantity of rice straw left over by the combines. Paddy straw is the stem part of the rice plant that remains in the paddy after the head has been harvested. It is generated at the tune of about 1.0-1.5 tonnes per tonne of paddy harvested. Paddy straw is relatively low in cellulose and high in hemi-cellulose content. It contains 15-30% silica on a dry weight basis. Because of its chemical composition, paddy straw in its natural form is not a popular animal feed because of its low digestibility, poor palatability, low protein (less than 5% in dry matter), high lignin and silica content, which are nutritionally inert and cause abrasiveness of gastrointestinal tract of cattle.

In winter burning of paddy straw results in severe environmental pollution in the northern India resulting in low visibility during dusk and dawn hours especially. About 5% of paddy straw is used as a cattle feed, 2% for making farm structures, 5% as raw material for making paper and card board, 7% as packaging material for horticultural crops and other industrial goods like china ware, etc and rest 80% is burnt in the field which causes nuisance to the environment resulting in unhealthy atmosphere having effect on human being, animals, crops and other living objects. It is important to realize that the traditional use of crop residues is not likely to increase much and the additional residues generated as a result of higher agricultural production will be entirely surplus to the traditional use (Singh, 2002, 2003, 2004). Utilization of the growing surplus of crop residue as energy source will have the following advantages:

a) Indian agriculture will have its own energy source, which will insulate it from future uncertainties in the supply of fossil fuel energy.

b) Surplus residue will become a saleable commodity, which will add to the income of the farmer.

c) It will decentralize power generation and supply giving more says to the rural community in the matter.

d) New jobs will be created. A five MW Crop-Residue based power project will generate at least one hundred thousand man days of employment per year in the rural sector.

e) Crop residue based dedicated power systems offering assured supply of energy would open the way for rural industrialization, which will generate more income and employment in the rural sector. The supply of grid power to rural sector is erratic and inadequate. This has blocked the development of industry including agro-processing in rural areas.

f) The practice of burning crop residue in the field, which causes serious air pollution in some regions, will stop.

A flail type paddy straw chopper-cum-loader (Fig. 9.5) has been developed at PAU Ludhiana (Singh, 2002, 2003, 2004). The machine cuts and chops the paddy straw in a single operation before filling in the trolley. The width of machine is 1.5 m. The machine is operated by any tractor 35 hp and above at standard PTO rpm of 540. The rotational speed of flails is about 100 rpm. The flails cut straw and throw it on screw auger. The auger is operated at about 320 rpm and it feeds the straw to the chaff cutter. The cutter cuts the paddy straw into small pieces, which is operated at about 1120 rpm. The cutter blades and blowing blades throws the chopped material to the trolley through long and high spout. The material loaded in trolley can be transported to the cardboard making factory.

Fig. 9.5: Tractor operated straw chopper-cum-loader in operation

Paddy straw spreader: an attachment to combine harvester

The combine harvester harvests the paddy crop from about 3.6 m width and throws the straw in about 1 m width. This improper distribution and spreading of paddy straw provides non-uniform feed to the subsequent operations like stubble shaving, straw chopping, disc harrowing, mould board ploughing etc. A paddy straw spreading attachment (Fig. 9.6) to combine harvester has been designed and developed at PAU Ludhiana (Singh, 2004). The paddy combine harvester is tractor operated. The tractor is mounted on the combine chassis. The power from the tractor PTO is transmitted in perpendicular direction with bevel type gearbox to the main shaft, which runs at about 900 rpm. Further power from main shaft is taken with the help of V-belt and pulleys to horizontal shaft mounted on the rear end of the combine. Then power is transmitted downward through vertical shaft with the help of bevel gears. The spreader consists of 4 blades of 30 cm length mounted on this shaft in horizontal plane, which spreads the paddy straw. The blades are in the form of straight strips attached to vertical shaft with the help of hollow circular pipes. The diameter of this spreader is 115 cm. The spreader is operated at rotational speed of about 300-400 rpm. Thus, spreader cuts the paddy straw with impact and spreads the paddy straw along swath of cut of combine harvester.

Fig. 9.6: A view of paddy straw spreader, an attachment to combine harvester

Otherwise labour is employed to spread paddy straw, which takes about 5 man-h/ha. In spreading operation, there is 3-6% cutting of paddy straw, which helps in reducing the size of straw. In this operation, dry leaves of paddy get converted into dust so whatever the weight of paddy straw is reduced that contributes to organic matter of the soil. This operation requires about 3-5% additional power for spreading of straw from power source of the combine harvester. This implement is quite useful for spreading paddy straw and provides uniform feed rate to the chopper or any other tillage implements for subsequent operations.

Straw Baler

Removal of paddy straw from the combine-harvested paddy field is a laborious and time-consuming job. Also, the straw needs a lot of space for storage. Straw baler is a machine which can pick the loose straw from the field (Fig. 9.7), compact it, tie into rectangular/round shape and cut into specified size so that bales can be easily transported to the desired point. Balers are classified in several ways like round or rectangular balers based on the shape of bale; field or stationary balers based on mobility; tractor mounted or self-propelled field balers based on mode of mobility; and manual, semi-automatic or automatic balers based on feeding, tying and bale-length controlling mechanism (Anonymous, 2010, 2012). The round balers are essentially automatic field balers while rectangular baler could be of any type. Balers use whole straw and no pre-processing is needed. The rectangular balers are essentially plunger type balers. The field balers essentially has a unit to either pick-up windrow and elevate, collect and elevate; a conveyor to move straw to bale-chamber entry; a packer to place straw in the chamber while plunger is on its retracted stroke; a reciprocating plunger to compress straw and move it through the bale chamber; means for applying force to resist movement of straw through bale chamber thereby controlling degree of compaction and bale density; a metering device for controlling bale length; means for separating bales and placing wire or string around the bale; and tying and cutting device.

Fig. 9.7: Straw baler in operation in combine-harvested paddy field
Courtesy: CLAAS India Ltd.

Generally stationary balers are not provided with first two components. Instead of first two components, it has a feeding chute for manual feeding. As plunger moves on its compression stroke, the straw in the bale chamber is compressed until the plunger force becomes large enough to move compressed mass through chamber. On return stroke, the compressed straw is held by spring loaded dogs that project into bale chamber and new straw is taken in. The energy requirements vary widely from 0.57 to 1.05 kWh/tonne. Generally, balers require a tractor of 40 hp or more.

There are two main types of balers. One makes small rectangular bales, weighing less than 25 kg. The other makes large bales, up to 900 kg, which may be rectangular or cylinder-shaped (round bales). Both big and small balers are pick-up balers but the term pick-up baler usually refers to the machine which makes small bales. The other type is known as big baler. A plunger-type field baler (Fig. 9.8) includes the following functional components (Sandhya *et al*., 2007):

- A unit to pick up hay from the windrow and elevate it.
- A conveyor to move the hay to the bale-chamber entry.
- Packers to place the hay in the chamber while the plunger is on its retracted stroke.
- A reciprocating plunger to compress the hay and move it through the bale chamber.

- Means for applying forces to resist the movement of hay through the bale chamber and thus control the degree of hay compression and the resultant bale density.
- An automatic metering device for controlling bale length.
- A means of separating consecutive bales and placing the wires or strings around each bale.
- Automatic tying device that operate when the bale reaches the preselected length.

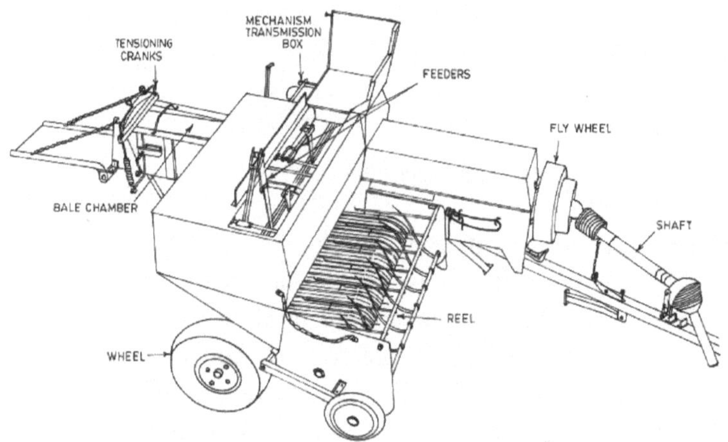

Fig. 9.8: Schematic view of straw baler

The baler has three major operating units. One unit is for picking the windrow or straw from the field, and then it is moved up in compression unit where the straw is compacted with plunger at a speed of 540 to 560 rpm in bales of rectangular shapes. Then, a knotter unit is provided to wind up the bales tightly (Fig. 9.9). It can form bales of varying length from 40 to 110 cm. The height and width of bales are generally fixed at 46 cm. The weight of bales varies from 15 to 45 kg depending on moisture content of straw and length of bales. The field capacity of baler for bale size of 60 x 48 x 38 cm is 0.80 – 1.0 ha/h. The weight of compressed tied bales for paddy and wheat straw is around 10-12 kg and 30-36 kg, respectively. The bale output and straw pick up efficiency are 200-250 per ha and 80-90%, respectively. Bales are transported to end users through trailers (Fig. 9.10). Compaction ratio for bales of paddy and wheat straw is calculated by dividing bale density by loose straw density. In case of paddy straw, compaction ratio is found to be 7 to 10 while in case of wheat straw it is 3 to 5.

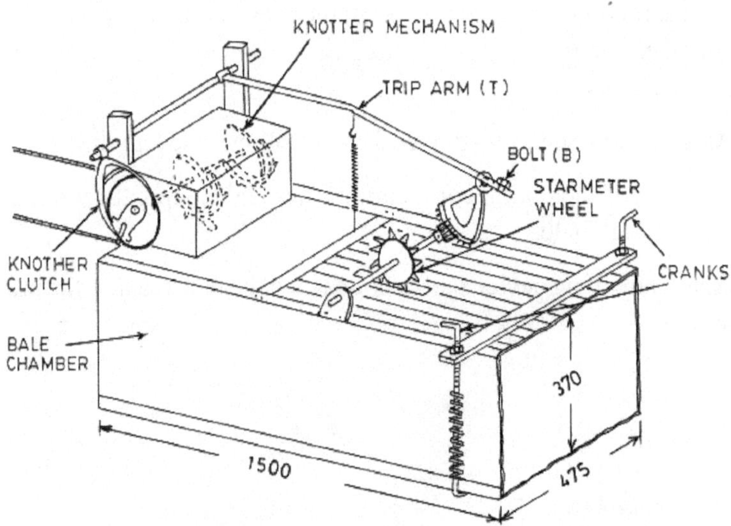

Fig. 9.9: Mechanism for varying bale length

Storage of harvested straw: The unbaled paddy straw collected for storage in developing countries is stored using two systems. One involves stacking straw in a large clump using a central pole of about 4-6 m height. These stacks are not protected by a roof, and consequently at the top of the hip the straw is stacked at an angle to facilitate the runoff rain. Most enterprising farmers build these stacks under trees, in a few instances; they may even be on raised wooden platforms or covered with polythene and corrugated iron sheets. These practices assist in maintenance of straw quality by reducing the leaching effect of the rainfall and microbial invasion that occurs when the straw is wet. The second method is to place straw in stacks under a roof to reduce the effects of rainfall. But this method is not common simply because of the capital cost of erecting building and the fact that paddy straw being bulky requires considerable space. The choice between these two methods depends on the resources available, the economic standing of the

Fig. 9.10: Transportation of bales

farmer and prevailing environmental conditions of which the rainfall is most important. The problem of bulkiness is overcome by baling and densification. These bales are neatly stacked, reducing the space required for the storage, and the stacks are kept either in open or undercover. The storage space requirements are directly related to the density of bales and method of stacking.

Tractor operated straw reaper (Straw combine)

The grain combines were introduced in India in the 1970's. The use of combines helped in timely harvesting of wheat but, it resulted in loss of 'bhusa' (cattle feed). In majority of cases straw left in the field is burnt. This practice not only leads to environmental pollution but also causes a considerable economic loss of precious biomass widely used as a cattle feed. The wheat straw is a valuable by-product. The wheat straw is fed to the cattle in the form of 'bhusa' i.e. finely cut and crushed. To collect and bruise the wheat straw and stubbles left behind after use of grain combines, a wheat straw reaper (combine) has been developed (Singh and Pandey, 2008; Verma *et al.,* 1992; Ahuja *et al.*, 1993). A straw reaper is a useful machine to recover the wheat straw, which is normally left in the field by a combine harvester. The straw combines cut the left over wheat stubbles and also collect the left over straw and ear-head bearing plants from the combine harvested field and do the threshing. Thus, it recovers the grain and loads the prepared 'bhusa' into the attached trailer.

A straw reaper essentially consists of four main units' viz., stubble cutting and collecting unit, feeding unit, straw bruising unit and '*bhusa*' blowing unit (Fig. 9.11), Garg and Singh, 2002. Several types of bruising drums are currently in practice. These include spike tooth, chaff cutter and serrated saw type. However, serrated saw, though costly is considered as most efficient straw bruising system (Ahuja *et al.*, 1993). Serrated saw type cylinder are mostly used in the straw combines for bruising. Serrated plates are attached on the bars at specific spacing and the bars are arranged parallel to drum axis. This machine cuts the left over straw at an average height of 7-8 cm above the ground, conveys it to bruising drum, where it is chopped/bruised into fine straw then blown into attached trolley. Wheat straw combines have been equipped with conventional harvesting combine cutter bar and conveying unit. Normally the cutter bar width is about 1.8 to 2.0 m. The reel in this straw combine is usually of a smaller diameter in comparison to grain combine. Currently available units are mostly drawn types (i.e. tractor in front). During machine operation the tractor tires trample some straw. Feeder conveyer feeds the straw to bruising cylinder. Identical to threshing drum in threshers, in this case a bruising drum is provided to bruise/chop the straw into fine pieces. Bruising unit has a closed-ended concave with openings of 10-12.5 mm size. Thus in this system the straw is not allowed to escape from the top of the concave. Wheat straw passing through the concave after bruising has been suitably reduced in size, to be used as chaff-straw. Bruising drum diameter and length varies between 60 to 90 cm and 100 to 150 cm respectively. Cylinder rpm varies from 500 to 750 rpm or a tip speed of 24 to 27 m/s. Material being fed to the bruising unit also contains few grains mostly on account of cutter bar losses. Material passing through the concave falls on a reciprocating sieve. Chaff and lighter materials are sucked by the aspiration fan and are delivered to the attached trolley through an adjustable blower duct. Aspirator normally has 3 to 6 blades with outer diameter ranging from 50 to 80 cms. Fan tip speed varies from 30-50 m/s (800-1000 rpm).

Straw reaper is pulled by tractor of 45 hp (34 kW) with an attached trolley. As soon as this trolley is completely filled with straw, it is unloaded near the dumping site normally located centrally or in the corner of the field or another trolley is attached and filled trolley is taken to dumping site. The field capacity of the machine varies from 0.4 to 0.6 ha/h at forward speed of 2.5-3.0 km/h. The straw recovery is about 55-60%. Straw recovery mainly depends upon the stubbles height remaining in the field after harvesting by combine harvester. Straw recovery rate varies from 25 to 35 q/ha. The grain recovery varies from 40 to 60 per cent. The grain collected in pan ranges from 50-100 kg/ha. Two persons are required for its operation.

Conventional tractor trolley is used as a transient storage during harvesting operation. This trolley is covered with a fine net, so that air is blown off and bruised chaff is stored in this. In order to maintain a continuous field operation of straw bruising unit, two to three trolleys have to be netted (or canvassed) and employed in this operation. This is essential because while the first trolley is being unloaded the second is used for straw combining. Number of trolleys to be used depends upon the transportation distance. One wheat straw combining unit and three transport trolleys and two tractors (one of 45 hp and another of 35 hp) are being used by the farmers. As soon as this trolley is completely filled with straw, it is detached and the second trolley is attached. This second trolley unit is made available close to the spot where the first trolley is being filled. Thus, straw combine unit is put to operation after attaching the new trolley whereas the filled trolley is detached and taken near the dumping site. Third trolley, which had been emptied by then, is attached and taken near the straw combine unit for refilling. The tractor use per year has increased with the use of straw combine. There is saving of time, money and labour. 'Bhusa' can be obtained as by-product in less time and less fatigue than manual threshing of wheat. Machine repays its cost within two years if a farmer uses the machine on custom hiring. The tractor needs frequent service of air cleaner because of dust and 'bhusa' during operation of straw combine. Straw reaper is very economical and now days it is used extensively by the farmers because of high cost of 'bhusa'. It helps to save environmental hazards.

Fig. 9.11: Tractor operated straw reaper (combine) in operation.

Chopper type tynes for power tiller rotavator for sugarcane trash shredding

For sugarcane crop cultivation, sugarcane trash management is of vital importance since it is usually burnt in the fields because the cost of labour required in collection and transportation of the trash is prohibitive. The sugarcane trash can be used for making organic matter. Power tillers are used for different operations in sugarcane crop viz., stubble uprooting, interculturing, earthing up, transportation etc. Power tiller rotavator is equipped with L-shaped hatchet type tynes for trashy conditions for weed control under wetland condition. If these tynes are used for sugarcane trash shredding, entanglement of trash

Fig. 9.12: Chopper type tynes for power tiller rotavator for sugarcane trash shredding.

around the rotavator shaft is observed (Anonymous, 2008). The normal angle for L shaped tynes is $118^0 + /-2^0$ which is changed to 160^0 to 180^0 to make it chopper type tynes (Fig. 9.12). The length of shank portion i.e. hardness zone is 100 mm. Overall length of tyne ranges from 150 mm to 210 mm. The length of cutting edge portion ranges from 40 to 80 mm. The hardness in edge portion is 56+/- 3 HRC and 37 to 45 HRC in shank portion as per IS: 6690 - 1981. Effective field capacity of power tiller rotavator equipped with chopper type tynes is observed to be 0.065 to 0.085 ha/h with 66.6 to 82.3% field efficiency. Significant reduction in sugarcane trash size is observed.

Tractor mounted rotary field shredder for sugarcane

After the harvest of sugarcane, trash and crop residues are left in the field which needs to be shredded for easy incorporation. For this purpose, a tractor mounted rotary shredder has been developed at the TNAU, Coimbatore (Anonymous, 2006). For sugarcane grown in ridges and furrows, especially in ratoon crop, the unit can be operated with the tractor PTO drive without damage to the stock ratoon. The rotary field shredder consists of two gangs of rotary swinging cutters which can slide on the control shaft depending on the row spacing (Fig. 9.13). Speed reduction and power transmission is achieved through a set of V-groove pulleys and a gear box. The whole unit is mounted in a casing which is attached to the tractor with 3-point linkage. Total width of operation is 1.75 m and it has easy maneuverability in the ratoon crop of sugarcane. At the forward speed of operation of the machine at 1.63 km/h, its field capacity and field efficiency are found to be 0.37 ha/h and 68.6% respectively. A tractor operated sugarcane trash incorporator chops up dry leaves into small pieces and mix up in the soil. It adds organic manure thereby increasing soil fertility. About 10% of sugarcane biomass is trash which could be incorporated in the soil rather than burning it. Burning of sugarcane trash is prohibited and harmful for the environment. It has been found that trash application at the rate of 10 t/ha release 14 kg nitrogen, 6.3 kg phosphors and 32 kg potash.

Fig. 9.13: Tractor mounted rotary field shredder for sugarcane

Power tiller operated shredder cum *in situ* incorporator

After the picking of vegetables, standing stalk of the plants remains in the field which is pulled out manually. This process is slow, laborious and involves drudgery. TNAU, Coimbatore has developed a power tiller operated roto shredder cum *in situ* incorporator (Anonymous, 2006). It consists of shredder assembly, power transmission system; hitch frame and rotary tiller attachment (Fig. 9.14). The gear box and shredding blade assembly are fitted on the main frame. The shredding unit consists of two back swinging type of rotary blades obtaining power through a bevel gear box and variable speed pulleys, to achieve a speed of operation of 900-1200 rpm. A counter weight of 50 kg is attached to the rear of the power tiller to balance the machine, enabling easy maneuverability. Field tests of the machine have shown satisfactory shredding and incorporation in the soil. The field capacity and field efficiency are 0.08 ha/h and 67.89%, respectively.

Fig. 9.14: Power tiller operated shredder cum *in situ* incorporator

References

Ahuja S S; Kalkat H S; Sharma V K. 1993. A perspective on wheat straw combine-its field performance and economic evaluation. Paper presented (FPM-93-2-96) at XXVIII Annual Convention of ISAE held at CIAE Bhopal, 2-4 March, 1993

Anonymous. 1987-89. Biennial Report. Department of farm Power & Machinery, Punjab Agricultural University, Ludhiana.

Anonymous. 1999. Souvenir on Paddy Straw Management at Annual day of Punjab Chapter of ISAE, held at PAU, Ludhiana on February 26, 1999.

Anonymous. 2006. Research Highlight. AICRP on Farm Implements and Machinery, CIAE Bhopal. *Technical Bulletin No.*: CIAE/FIM/2006.

Anonymous. 2008. Research Highlight. AICRP on Farm Implements and Machinery, CIAE Bhopal. *Technical Bulletin No.*: CIAE/FIM/2008/141.

Anonymous. 2009. Technical Folder. AICRP on Farm Implements and Machinery, CIAE Bhopal. *Technical Bulletin No.*: CIAE/FIM/2009/144.

Anonymous. 2010. Research Highlight. AICRP on Farm Implements and Machinery, CIAE Bhopal. *Technical Bulletin No.*: CIAE/FIM/2010/151.

Anonymous. 2012. Directory of Successful Farm Machinery in SAARC Countries. SAARC Agriculture Centre. BARC Complex, Farmgate, Dhaka – 1215 (Bangladesh).

Garg I K; Singh Surendra. 2002. Farm equipment for Punjab agriculture. Department of Farm Power & Machinery, Punjab Agricultural University, Ludhiana.

Pandey M M; Majumdar K L; Singh Gyanendra; Singh Gajendra. 1997. Farm Machinery Research Digest. Technical Bulletin No. CIAE/97/69, Central Institute of Agricultural Engineering, Bhopal, 328 p.

Sandhya; Singh Surendra; Dixit Anoop. 2007. Performance evaluation of field balers. *J. of Agril. Engg.* Vol. 44(1): 43-47.

Singh Surendra. 2002. Annual report of Mechanization of rice-wheat cropping system for increasing the productivity (NATP-project), Department of Farm Power and Machinery, PAU, Ludhiana.

Singh Surendra. 2003. Annual Report of NATP scheme on 'Mechanization of Rice-wheat cropping system for increasing the productivity'. Department of Farm Power & Machinery, P.A.U. Ludhiana.

Singh Surendra. 2004. Annual Report of NATP scheme on 'Mechanization of Rice-wheat cropping system for increasing the productivity'. Department of Farm Power & Machinery, P.A.U. Ludhiana.

Singh Surendra. 2007. Farm Machinery – Principles and Applications. Directorate of Information & Publication of Agriculture, Indian Council of Agricultural Research, Krishi Anusandhan Bhawan-I, Pusa Campus, New Delhi.

Surendra Singh; Pandey M M. 2008. X Plan Achievements (2002-2007). AICRP on Farm Implements and Machinery, CIAE Bhopal. Technical Bulletin No.: CIAE/2008/137.

Verma S R; Kalkat H S; Singh Joginder. 1992. Straw combine-A new development in Agril. Machinery Agril. *Engineering Today, ISAE* 15-16 (1-6):24-31.

Zeitfracht Medien GmbH
Ferdinand-Jühlke-Straße 7
99095 Erfurt, Deutschland
produktsicherheit@kolibri360.de